Tolley's Fire Safety Training Manual

by
Dan Sullivan
Lawrence Webster Forrest

Members of the LexisNexis Group worldwide

United Kingdom	LexisNexis UK, a Division of Reed Elsevier (UK) Ltd, 2 Addiscombe Road, CROYDON CR9 5AF
Argentina	LexisNexis Argentina, BUENOS AIRES
Australia	LexisNexis Butterworths, CHATSWOOD, New South Wales
Austria	LexisNexis Verlag ARD Orac GmbH & Co KG, VIENNA
Canada	LexisNexis Butterworths, MARKHAM, Ontario
Chile	LexisNexis Chile Ltda, SANTIAGO DE CHILE
Czech Republic	Nakladatelství Orac sro, PRAGUE
France	Editions du Juris-Classeur SA, PARIS
Germany	LexisNexis Deutschland GmbH, FRANKFURT and MUNSTER
Hong Kong	LexisNexis Butterworths, HONG KONG
Hungary	HVG-Orac, BUDAPEST
India	LexisNexis Butterworths, NEW DELHI
Ireland	Butterworths (Ireland) Ltd, DUBLIN
Italy	Giuffrè Editore, MILAN
Malaysia	Malayan Law Journal Sdn Bhd, KUALA LUMPUR
New Zealand	LexisNexis Butterworths, WELLINGTON
Poland	Wydawnictwo Prawnicze LexisNexis, WARSAW
Singapore	LexisNexis Butterworths, SINGAPORE
South Africa	LexisNexis Butterworths, Durban
Switzerland	Stämpfli Verlag AG, BERNE
USA	LexisNexis, DAYTON, Ohio

First published in 2003
© Reed Elsevier (UK) Ltd 2003

A CIP Catalogue record for this book is available from the British Library.

ISBN 0 7545 2183 4

Typeset by Phoenix Photosetting, Chatham, Kent
Printed and bound in Great Britain by Antony Rowe Ltd, Chippenham, Wilts
Visit LexisNexis UK at www.lexisnexis.co.uk

About the Author

Dan Sullivan

Dan Sullivan is a very experienced and well-qualified fire safety professional who served for 25 years in local authority fire brigades, some ten of those years as a fire safety inspecting officer. With this thorough grounding in the subject – and a subsequent career with fire engineering and risk management consultants, Lawrence Webster Forrest – Dan has gone on to acquire an enviable portfolio of training and communication achievements, including a Postgraduate Certificate in Education and many attainments in the field of public speaking (he is an active member of Toastmasters International).

This passion for communicating has led to the development and presentation of a huge range of fire safety courses – again in both the public and private sectors – and success in meeting the training needs of employers and their workforces all over the country. Dan has written for professional journals and magazines, worked on community education campaigns and hosted national and international conferences for both business and leisure purposes.

Contents

Contents

Introduction to the Fire Safety Training Manual

Overview

Today's manager with delegated responsibility for fire safety in the workplace is confronted with different pieces of relevant legislation that can appear complex and which has a tendency to change on a regular basis. Indeed, fire safety legislation is currently undergoing the most dynamic period of change since the early 1970s. These changes have implications for the organisation and to the individuals who accept delegated responsibility.

This publication aims to demystify fire safety law and define where fire safety training fits within the overall fire safety responsibility of organisations. It also aims to provide options for fire safety training and to propose a system of training staff at different levels within the organisation with different levels of responsibility. Recent changes in legislation with respect to fire safety, and health and safety generally, are placing the onus of responsibility for risk definition and measures to offset the identified risks with the employer. This concept is intended to drive safety awareness down through an organisation, from the top management where responsibility ultimately lies, to the staff charged with implementing strategic policy.

Use of the Fire Safety Training Manual

The publication is in three parts:

The first part introduces the general approach to fire safety training by discussing current and future trends in fire safety law, the evaluation of training objectives and where the function of training fits within the fire safety strategy of the organisation. This part also develops what options there are for training and what the advantages and disadvantages of each are. Having clarified why fire safety training must be delivered, the content of courses is discussed and we consider who delivers the training package.

The second part provides standard fire safety training modules with supporting trainer's notes. Further specific occupancy risk modules are also included where the workplace is acknowledged as having risks peculiar to the use and occupancy of the buildings.

The third part provides further training notes and guidance for the trainer with advice on how to demonstrate fire safety competency within the workforce. This includes a range of questions to use in post-course questionnaires to enable competency judgements to be assessed for the different grades of delegated responsibility. A selection of FAQs (see **Part 3**) is also presented drawing on the author's experience in the preparation and delivery of fire safety training in a multitude of work environments.

The publication therefore acts as a trainer's guide explaining why training is important, how to train, what the training options are, who trains and when training should take place. The second part of the publication provides sufficient information and a format for the provision of training courses for the workplace.

Part 1: The General Approach to Fire Safety Training

Statutory Duty

Many pieces of current legislation are relevant to heath and safety and fire safety in the workplace. The following Acts and Regulations are considered here, as they affect the requirement and content of fire safety training.

A duty is given to employers under *section 2* of the *Health and Safety at Work etc Act 1974 (HSWA 1974)* to provide training to ensure, so far as is reasonably practicable, the health and safety of employees in the workplace. This Act imposes general requirements on the employer to provide a safe working environment. It recognises through empirical evidence and research that a key element in safety is the definition of safety procedures, which in turn are related to an organisation's staff through a training regime.

Two pieces of legislation relate specifically to fire safety and are the principal instruments by which fire safety is defined and enforced in the United Kingdom.

The *Fire Precautions Act 1971 (FPA 1971)* (as amended) is the responsibility of the Home Office and is enforced through the local Fire Authorities (Northern Ireland has an equivalent Act and enforcement procedure). The *FPA 1971* enables the fire authorities to require fire safety training and this is often reflected in conditions issued with a fire certificate. The specification for fire safety training, content and frequency within a fire certificate makes the requirement a statutory duty on the employer.

The *Fire Precautions (Workplace) Regulations 1997 (SI 1997/1840)* (amended 1999) are based on a self-cognisant approach in which the employer is responsible for identifying fire risks and implementing measures to mitigate them. The Regulations also require that employers 'provide information and training to your employees about the fire precautions in your workplace'. This training will obviously be required to respond to specific risks within the workplace as identified within the fire risk assessment.

Potential Changes to Fire Safety Legislation

Governments continuously align themselves with the objective of simplifying legislation in all its forms. This approach has applied to fire safety legislation in the past and is again a current proposal. The *Fire Precautions Act 1971* came about following a review of the then disparate state of fire safety legislation. In excess of 60 separate pieces of legislation were in force at that time and the administration and enforcement of the various Acts was complex and unwieldy. The intention of the *FPA 1971* Act was to consolidate this plethora of regulation into a single Act administered through a single authority.

By and large the *FPA 1971* was successful in consolidating the majority of legislation. However, it has many shortcomings, perhaps the greatest of which is the restriction in scope to certain premises and its disregard to the area of occupation where most life is lost, in the home.

New legislation was introduced by the current Government to help simplify law rationalisation. It is an Act which itself enables other legislation to be simplified without the lengthy and cumbersome process of repeal and enactment otherwise necessary. This legislation is known as the *Regulatory Reform Act 2001*. It is under this Act that current changes to fire safety laws are proposed. The Office of the Deputy Prime Minister has issued a consultation document outlining the intention to consolidate further the current position.

Referring to the two principal pieces of current legislation discussed above, it is evident that the two statutes apply, in part to the same premises. For example, the *FPA 1971* applies to offices, shops and railway premises, for which under certain conditions a fire certificate is required. The *Fire Precautions (Workplace) Regulations 1997 (SI 1997/1840)* also apply to these premises. This position is obviously confusing to building occupiers, as it is a fact that the enforcing authority has different powers to make requirements under each statute.

This unfortunate position was to some extent brought about by the Governments duty under European Directives to align fire safety with health and safety legislation, to encourage employer participation in risk management and to devolve safety awareness down through an employer's staff. The general idea here was to raise fire safety awareness in the employed population through participation and awareness leading to self-risk management and administration. The Government in its first attempt to introduce the EU Directive, created the *Fire Precautions (Workplace) Regulations 1997*. This original version only applied to those premises not currently benefiting from a fire certificate issued under the *FPA 1971*. The requirements of the Directive were therefore only imposed on those places of employment falling outside of the requirement of a fire certificate. On review of this situation the EU found that the requirements of the Directive had not been met and the British Government had to subsequently amend the Regulations in 1999 to include all places in which people were employed.

The current proposal and main thrust of the consultative document issued by the Office of the Deputy Prime Minister is to replace much of existing and conflicting legislation including the *FPA 1971* and the *Fire Precautions (Workplace) Regulations 1997* with a new piece of legislation. The new statute will be based on the same concept of self-risk management as the current *Fire Precautions (Workplace) Regulations 1997*. It will apply (with very minor exceptions) to all places where people work, access and congregate.

The new legislation will be based on self assessment and management of fire risk and the emphasis will be on the need (to a greater or lesser extent) to rely on fire safety procedures to offset otherwise unfavourable risk conditions imposed by potential limitations of the building.

Even in a building fully equipped with comprehensive physical fire precautions a basic procedure for the evacuation of the premises in the event of fire will be necessary. Any procedures, and indeed any fire precautions installations, are only effective if the people they are designed to protect are aware of what to do, and how to respond to the signals and information given by them.

Assessment of Training Objectives

Fire safety within any organisation must be approached in a systematic way. The provision of a safe place for employees and visitors to work and resort does not happen by accident, it is a function of design. Fire precautions, the generic term for fire safety measures and the legal system that defines and enforces fire safety, encompasses physical provision within a building as well as the management intervention necessary to ensure life safety.

The focus of current legislation is on the protection of life. No real consideration is given within any Act or Regulation towards the protection of property from fire over and above that level which is provided by default when protecting occupants. An employer will have to invest his own energies into setting appropriate standards of protection against fire for his business processes, property contents and building fabric. It is a matter of statistical record that, of those businesses that experience a serious fire which prevents normal trading, less than half survive beyond three years of re-commencement of business. This is attributed to the migration of regular business customers to competitors. The fact that insurances are in place for collateral damage and business continuation protection may be irrelevant: the market has moved on in a way that has proven fatal to the business.

Figure 1 below defines the holistic fire precautions system. You will note that 'all paths lead to and from the Fire Risk Assessment' and this is because it should be a central tool to the person with fire safety responsibility in an organisation.

Fire Safety Policy and Procedures

The system represented by **Figure 1** defines the three strands of fire safety in the workplace:

- Passive Fire Precautions (structural and physical protection from fire);

- Active Fire Precautions (fire safety and fire fighting systems installations); and

- Management of Fire Precautions (management and staff roles and responsibilities)

It is not the scope of this publication to consider the appropriateness of the passive or active fire precautions in your workplace. It is assumed that your fire risk assessment process has evaluated the contribution each element of the physical installations has made to risk management. Indeed, it is the role of the fire risk assessment to identify any residual risk and to define either physical or fire safety management procedures to offset the identified gaps in the safety management process.

Here we will concentrate on the 'Management of Fire Precautions' strand given in **Figure 1**. Four contributory parts are identified: Policy, Procedure, Training and Drills. Each is discussed and outlined to give further context to the principal issue of training.

Policy

All companies should have, and regularly review, a fire safety policy document, no matter what size organisation and how many people are involved with the enterprise. The Policy document should contain a statement giving clear commitment of the senior management/proprietor of the organisation to comply with all current fire safety legislation. It should state its interpretation of which legislation is applicable to the organisation and how it intends to meet the requirements set by the Acts and Regulations.

The Policy document should also set out in terms of the organisational structure of the company, who has ultimate responsibility for fire safety. This may be the company chairperson, chief executive or managing director. In the case of other forms of business entity the ultimate responsibility may lie with a Board of Trustees or Governors dependant upon the legal entity and status

Figure 1: Fire safety in the workplace

of the organisation. Sometimes these governing bodies are legally disassociated from the day-to-day management, particularly if the organisation is a 'not for profit' trust or other formation that occupies and manages buildings to which people have access. In these cases the Policy document will define the relationship between ownership/stewardship and day-to-day management and responsibility for the fire safety of employees and others coming into contact with the operation of the organisation.

The Policy document will define the role of any fire safety committee set up with the intention of communicating fire safety risk management issues between the senior management and employees and representatives of Unions and general staff. The Policy document will make commitments by senior management to make available sufficient resources, both human and financial to implement fire safety requirements identified within the fire risk assessment process and to maintain fire safety awareness throughout the body of staff as necessary.

The Policy document will accept that senior management of larger organisations cannot undertake all the functions necessary to maintain a satisfactory level of fire safety in the workplace themselves. The document will define devolved responsibility down through the hierarchy of the organisation. It will state what responsibilities are levelled with each post and it will make commitments to ensure provision of training is adequate to meet these responsibilities.

In summary, the Fire Safety Policy document states the commitment of the organisation to meet its obligations in law, to identify the hierarchy of responsibility for fire safety and to set out the means by which the delegated responsibility will be actioned. The document will be reviewed regularly (at least annually) to ensure that conditions, circumstances and statutory duties remain unchanged. The document will be endorsed by the most senior person in the organisation's management.

Procedure

Procedures are necessary to codify responsibilities into appropriate actions in response (in this case) to an emergency situation. Fire safety procedures are statements that define actions that each person with a delegated responsibility should respond to under certain stimuli. Procedures may also form part of a preventative regime to control fire risks.

In general, and for larger organisations, procedures are defined for differing levels of responsibility. For example, fire safety wardens or marshals may be appointed with specific duties. An individual may be charged with taking control of an incident from Fire Control Centre or given duties to take a roll call at the Assembly Point. Each specific responsibility in this case should be set out at least in outline and not be left to the interpretation of the individual. The reason for this is that the fire safety systems and available means of escape are fixed and specific to that building. The procedures will take these into account and respond to the fire safety design of the building.

In this respect buildings can differ greatly. Some will evacuate immediately on fire alarm activation, some will evacuate progressively, some systems will delay evacuation under controlled conditions before sounding an alarm, some public buildings will have discrete alarms that only members of staff will be aware of. All of these and other complications such as multiple occupancy buildings underline the necessity of detailed and co-ordinated fire safety procedures for all but the smallest of buildings.

Fire Action Notices giving standard instructions relating to actions on discovering fire or hearing the fire alarm are the cause of a common misconception. It is often thought that the directions given on these Fire Action Notices posted at fire alarm call points within a building constitute the fire safety procedures for a building. This is not the case for all but the smallest of premises. For all the reasons given above detailed fire safety procedures form the link between physical features and facilities within a building provided for life safety, and the actions that all should take in the event of an emergency fire condition.

Training

Fire safety installations such as fire alarm systems are devices that provide information to building occupiers. Fire alarm systems, unlike fire suppression systems, do not fight fire or intervene in the development of fire; they merely advise the occupants that a problem exists. Fire alarm systems that incorporate automatic fire detection provide, however, one of the most effective means to effect a safe occupancy and such installations are often able to compensate for deficiencies in other more traditional fire precautions. This reliance places additional importance on procedure definition.

Fire safety procedures are only effective if the staff or persons charged with emergency response are aware of the system's capabilities and limitations. In this sense it is imperative that a fire safety training regime is designed and delivered to all persons with delegated responsibility, and to the general staff, so that they know what to do in an emergency. It would be unrealistic and unfair for an organisation to delegate responsibilities for fire safety to people without providing the training to enable a level of competency to be achieved in the role. Indeed, it would be within the delegated person's rights to insist that acceptance of any such responsibility would be conditional on receiving adequate training.

Fire prevention, together with the identification and control of fire risk, is just as important as response to emergencies. Fire safety training should provide a level of awareness amongst all building occupants such that it becomes second nature to identify and resolve/report fire risk issues.

Fire safety training should therefore consist of two distinct but equally important parts:

- general matters – the law, fire safety housekeeping, fire science and fire fighting and general fire precautions; and
- specific matters – the precautions defined in the procedures prepared for the relevant building.

It is a common failing that organisations buy-in or deliver non-specific 'off the shelf' fire safety training packages which may address the general matters but ignore the very important aspects of evacuation and response peculiar to the premises in question.

Drills

In much the same way that procedures are pointless without appropriate training to ensure familiarity, training is only effective if the procedure for escape is practised regularly. The fire drill is the means by which those with specific responsibilities for fire safety actions in an emergency, and indeed general staff, can practise and become familiar with the process of evacuation under differing conditions.

Fire evacuation drills should be carried out at least once a year and preferably twice. The frequency of fire drills may vary from this, depending on the identified risk of the premises. If a working environment has high-risk processes or the form and layout of the accommodation is subject to change, or the staff turnover is high, fire drills may be required more frequently still and this should be identified within the fire risk assessment process.

It is preferable that drills are varied. If the building has several exit routes from lower and upper levels, it is a good idea to simulate an emergency condition and make one of the routes unavailable. In this way people will be encouraged to use exits less familiar than their normal access and egress routes. This process will encourage wider familiarity with the building layout, which in smoke-filled conditions could be a life saver.

Fire drills will often show up weaknesses in the procedures or in the response of staff and visitors to the premises. The process should be recorded and timed to ensure effectiveness. The person with responsibility for fire safety within an organisation should review the records of drills and look for improvements, and areas of the process that worked less well. The analysis could lead to changes in the written procedures, improved training courses, or a combination of measures to achieve optimum response.

Summary of Training Objectives

On completion of training at different levels as necessary, the trainees should be able to:

- understand fire safety law and how it applies to the organisation and to the individual personally;
- identify with and understand the strategy put in place by the employer to ensure fire safety;
- understand the individual's delegated responsibilities and duties under the imposed strategy;
- understand basic fire safety issues including risk, what fire is and how it is caused, and how best to mitigate and prevent fire occurring;
- identify with the specific fire safety evacuation and emergency response procedures for the building; and
- attain a predetermined level of competency to respond to delegated fire safety responsibility.

Training System

There are many options to consider when setting up and implementing a fire safety training system within an organisation. The 'training system' comprises of the definition of training requirement and the delivery of information to satisfy legal and moral duties. This section attempts to define training in terms of what is taught, who trains, resources and facilities. It also discusses what training delivery options exist.

Training System Options

What is Taught

As mentioned previously, the fire safety policy and related procedures will structure how certain people will respond in different conditions to the threat of fire. The procedures may stipulate a hierarchy of response and duty. For example, a building manager may be designated as an incident co-ordinator with central command responsibilities in terms of making judgement about evacuation, incident escalation, emergency services response, fire systems control and building reoccupation following an incident.

Building, area or departmental managers may be delegated responsibility for general fire safety housekeeping, the safety of disabled persons in their area, encouraging evacuation and ensuring people have left the building. Individual members of staff may have responsibility for ensuring that they work safely without putting others at risk and for their own safety in evacuation. Some staff may have specific duties such as Assembly Point recording officer (roll calling), or if a third party security service is employed for the building in question, liaison with the security managers.

Many multiple occupancy buildings and some owner occupied buildings rely on outsourced security or concierge services. Depending on the extent of the 'terms of reference' of the security/reception service, senior managers within the service provider should be included within the training system as they may perform a vital function in the emergency procedures.

The procedure's definition and the way in which responsibilities are designated within an organisation will therefore dictate the hierarchy of training. Clearly, those with specific and particular responses will require a different level of training from that of general staff members who have responsibility only for themselves in an evacuation situation. Recognising the need for all fire safety managers (fire marshals, fire wardens) to have deputies assigned to cover their responsibilities in times of absence, for smaller organisations particularly this may imply the provision of the same level of training to all staff. Larger organisations will choose a more selective approach to training particularly reflecting the nature of their business. This is one function of the Fire Risk Assessment process.

What to teach, and to whom are therefore a fundamental questions that need to be addressed at the outset. The answers will depend on the policy and procedures and the nature, availability and disposition of staff and other human resources capable of implementing them. This definition is the outcome of a 'training needs analysis'.

Who Trains

Before discussing the options for implementing fire safety training and rolling it out in the workplace, it is useful to consider trends in legislation as discussed above. Recent changes in the law including the introduction of the *Fire Precautions (Workplace) Regulations 1997* have stressed the move away from the imposition and policing of fire safety by the fire brigades and the refocus of responsibility onto the employer who is required to identify and rectify shortfalls in fire safety standards. One of the key objectives of this move, in line with general health and safety legislation, is to devolve safety awareness down through the organisation, and promote understanding and participation in all staff – in essence, to create a safety culture within organisations.

With this in mind, the wholesale outsourcing of fire safety training definition and delivery is not necessarily achieving this objective. With the idea of underpinning the fire safety culture within the organisation, we look at options for the development of training programmes and material and the subsequent delivery of the message to all staff.

Option 1 – Professional consultancy (total service)

This option often adopted by larger firms and organisations relies on the appointment of a fire safety consultant to provide a range of inclusive fire safety management services. These may include definition of Policy and Procedures and may link this developed strategy with the fire risk assessment process. The consultant will prepare a fire safety training regime that reflects the specific fire safety risks of the premises occupied by the organisation. The training programme will be based on an analysis of training need and reflect a system for judging competency of trainees to fulfil delegated roles.

The consultant will then deliver the courses to the staff at all levels of responsibility within the organisation. Courses will be provided initially and then at regular intervals as 'refresher courses'. The consultant will also train new employees as part of their induction into the working practices of the organisation. In brief, the consultant will identify strategy, design courses and deliver on an on-going basis training to all levels of the organisation's staff.

Advantages – This option will provide an integrated strategy for the organisation, reflect its requirements and be specific to the occupation, processes and risks the organisation faces. It will meet all statutory duties for training placed on the employer and will be delivered in a way that will ensure understanding. Feedback from staff can be professionally addressed and responded to immediately.

Disadvantages – The employer is being responsible in that the issues of fire safety training are being addressed and the employer is investing the financial resource to ensure all staff are trained in fire safety. There is however an element of abdication of responsibility. The implicit aim of current legislation is to devolve the fire safety responsibility through the organisation and to develop the safety culture in the workplace. As with any system imposed by a third party, it is rarely 'bought into' by the people whom it affects. There is the possibility that the consultant may try to impose a 'standard' solution adapted to the client's circumstances different from the normal business approach of the client organisation, and that may sit uncomfortably with staff and managers.

Option 2 – Professional consultancy (partial service)

Using this option the client may employ the consultant to work with managers to develop a realistic fire safety strategy with associated Procedures. The consultant may then develop a tailored fire safety-training course focused specifically on the training needs of the organisation. In this respect this option has the advantages of Option 1 but the delivery of training differs significantly. A system is devised that trains managers within the organisation to deliver the training down through the organisation. This approach could be termed 'training the trainer'.

The consultant may be employed on an on-going basis, but only to audit the system, its delivery and the resultant competency of the trainees.

Advantages – This option has the advantages of Option 1.

Disadvantages – Fewer disadvantages are obvious from this option except (as with Option 1) for the cost of consultant employment and the 'down-time' (loss of productivity) of the client staff involved in the development of the training regime. It must be accepted that training delivery will have an impact on time utilisation of all staff.

Option 3 – Professional training service

This option involves a professional training organisation coming to the premises of the organisation, or providing the training venue facility to deliver fire safety training on a regular basis.

The main difference between this option and that described for Options 1 and 2 is that the trainers deliver a largely standard training product rather than a developed regime focused on needs.

Advantages – The courses are readily available with a minimum of 'down-time' and can be provided on and off site.

Disadvantages – The courses often are very generic and not specific to the risks apparent in the employer's business. They can rely on video and other media that may be interesting the first time of watching but become tedious on repeated exposure. Without varied approaches to course development fire safety training can become repetitive and something that employees will seek to avoid, totally negating the prime purpose of the exercise.

Option 4 – Fire safety manager

A person is either specifically employed (having fire safety experience), or an existing member of staff, usually a senior manager, is charged with developing a fire safety system and associated training.

Advantages – The fire safety system is developed with the ownership of staff and there is usually a very positive impact on awareness and fire safety culture within the organisation.

Disadvantages – If the job of fire safety manager is a dedicated post employing a competent fire safety professional there are few disadvantages to this option. However, if the title 'fire safety manager' has been given to an individual in addition to a full job function, the position is very different. Pressures of everyday management often take preference over fire safety system development and training. Further, if the individual has not had a fire safety professional background, the level of competence and quality of system and devolved information may legitimately be questioned.

General comment

Four options are outlined above to indicate the main and most obvious ways in which to fulfil legal obligations for fire safety training. Many combinations and indeed totally different approaches are possible. It is intended here to highlight the primary advantages and disadvantages of certain approaches.

Fire safety training and fire precautions do have an impact on business resources and this is addressed separately below.

Training Resources

Human factors

When employing a third party to undertake fire safety training, whether meeting the service definition given above as 'consultant' or 'training service', it is essential to ensure the competency of the firm or individual. It is difficult to set guidelines in this respect except that there exist qualifications in respect of technical competency (fire precautions) and for delivery of training. As the individual, who may be delivering a set course could be questioned from the floor of the training room, it is important that their technical ability is such that questions can be competently responded to. Failure in this respect undermines the entire process.

The delivery of training should engage the audience and deliver the message in a way that relates to the individuals and is in no way ambiguous. Good trainers have often undergone training themselves and achieved status as an 'adult educator'. In any respect, the consultancy, training service provider or individual should be asked to outline their fire safety training experience. References from similar organisations to your own should be sought – this is often an enlightening exercise.

Competency

Assuming that the training system is developed and the quality of course content and trainer credentials are established, it remains that the message needs to be delivered and received in such a way that skills transfer occurs and a level of competency is devolved to the audience. It is a legal responsibility to provide fire safety training, however no specific guidelines are given with respect to judging the success of the training provided. Two issues are relevant here. The first relates to the organisation's duty to ensure that it has provided the tools (knowledge and physical resources) to an individual that it has devolved fire safety responsibilities to. The second is that an individual has every right to refuse a responsibility unless they have received competent training and feel comfortable in their own ability to discharge that responsibility.

How can the organisation judge fire safety competency across its delegated safety hierarchy? The system advocated here is the post-training course questionnaire. This approach is believed to have two primary advantages:

- it quantifies understanding; and

- when attendees are made aware of the potential of a questionnaire, it tends to focus attention during the course.

A disadvantage may be that some people may feel threatened by such a test and avoid the training by any inventive means they can think of.

From experience, reasons that persons avoid training with associated tests can range from general nervousness to weaknesses in literary skills. It is for this reason that when testing is applied it is done in a sensitive way and one that does not require an individual to write extensively, merely to tick option boxes. There is still a reading ability requirement and employers should consider this when designing training programmes as they are best placed to judge the general literary skills of the workforce.

Part 3 of this book provides typical test papers that respond to the generic training modules given in **Part 2**. These are intended to assist organisations that adopt the Option 3 approach given above and may be developed and varied to provide a range of relevant and testing questions. Further advice on the application of competency testing is provided in **Part 3** as is the development of training records and developing a system that can be audited.

When to deliver fire safety training

It is a legal requirement to ensure that all persons working for an organisation receive fire safety training and this should be repeated at least once a year. In some instances, and for buildings issued with a fire certificate under the *Fire*

Precautions Act 1971 as amended, a more frequent interval of fire safety training may have been required by the fire authority. An organisation's own fire risk assessment will consider frequency of training requirement. Factors such as fire risk levels, turnover of staff, or a very dynamic working environment will affect training frequency recommendations.

Another factor affecting frequency of training is the degree of responsibility devolved to an individual or group of individuals. Fire safety managers, fire marshals or wardens, or any person with specific duties with respect to fire should receive training twice a year.

Newly employed persons as part of their induction to the company should not be required to wait until the next scheduled full general staff training session, rather they should receive immediate basic induction information including, but not limited to:

- What action to take on discovering fire or hearing the fire alarm.

- Where all available means of escape are located in areas of buildings they are likely to have access to.

The entire training routine should apply to all grades of staff from Directors to temporary and part-time workers and is particularly important where people with disabilities are concerned. Again, the fire risk assessment process should be used to evaluate any special precautions necessary in the employment of, or access given to the disabled. Specific training is then required to reinforce any procedure devised for giving assistance to disabled staff and visitors.

All members of staff within an organisation should receive training to the frequency discussed above. No exceptions should be made at any level within the business. It is acknowledged that all organisations have critical functions and the staff involved in those functions are difficult to release for training purposes, so there is a clear need to plan for this situation in advance. The result of such planning is a spread of courses given at different times to allow cover to be retained in critical process areas and to cover for unsociable hour workers (night-time, weekend).

Part 2: Training Modules

To Help You Deliver Effective Fire Safety Training

Introduction

By careful use of the material provided here, a trainer with little previous knowledge of fire safety law, procedures or design, can quickly become familiar with the main concepts and requirements. Well-planned and progressive use of the modules will then enable the trainer to pass on the necessary level of understanding to ensure the safety of staff and others in the workplace.

While **Parts 1** and **3** of this manual give considerable background information and guidance on the delivery and monitoring of fire safety training, Part 2 forms the essential reference material, logically arranged and broken down into Modules. Legal and technical jargon has been largely eliminated, with all terms clearly explained in everyday language.

The Modules are divided into two categories: General (Modules **1–7**) and Specific (**8–12**). Within these categories, each Module starts with an introduction (or 'contents') followed by outline lists expanding upon each of the essential subjects covered within the relevant Module. On the right-hand pages, opposite the outline list is the corresponding explanatory text.

The Manual, and Part 2 in particular, is designed to facilitate in-house training based on a trainer delivering the message face-to-face in a suitable meeting, seminar or training room. **Part 3** provides considerable guidance on training techniques, but also stresses the importance of being familiar and comfortable with the training material in advance of any presentation.

Workplace Fire Safety

General Training Modules 1–7

Modules **1–7** in this section of the manual contain information that is relevant to almost every employer, and must be understood by employees. For this reason much of the content – particularly where it touches upon legislative, design or policy issues – has been simplified to the point where it serves more as an 'aide-mémoire' than an authoritative reference. The priority is to ensure a high standard of fire safety throughout the workplace, by teaching staff the fundamental principles of fire safety provision, good practice and legal compliance.

In every case the main points to be aware of are listed on the left-hand page, with the corresponding text on the right. In some areas we have included reference to other potential sources of information, should the need arise to study the subject in greater depth or detail.

Fire safety considerations more relevant to specific areas of business are covered in Modules **8–12**. It should not generally be necessary to study Modules unrelated to one's own field of employment.

Module 1:
FIRE SAFETY LAW

○ *The Fire Precautions Act 1971*

○ Contents of a Fire Certificate

○ The *Fire Precautions (Workplace) Regulations 1997 (SI 1997/1840)*

○ Building Regulations

○ Licensing and Registration Arrangements

○ Change on the Horizon

The *Fire Precautions Act 1971* was the first major piece of legislation dealing exclusively with fire safety matters, and took over the issuing of Fire Certificates, previously dealt with under various *Factories Acts* and *Offices, Shops and Railway Premises Acts*. In recent years the *Fire Precautions (Workplace) Regulations 1997* (as amended) have broadened the scope of fire safety legislation, and given employers greater responsibility for determining standards to be met, by requiring them to carry out a Fire Risk Assessment and act on its findings.

The Fire Brigade has considerable powers of entry associated with the enforcement of this legislation, and can insist upon being allowed to inspect most places of work at any reasonable time – this is usually interpreted as meaning whenever there is someone on the premises.

If it is considered that contraventions have been committed, the Fire Authority can impose requirements, or commence legal proceedings leading to prosecution. Where a prosecution is successful, courts can fine the occupier (or in some cases the owner) of the premises, or even an individual employee found to be at fault. These fines can be considerable.

1a: The Fire Precautions Act 1971

○ Deals mainly with Factories, Offices, Shops, Railway Premises and Hotels

○ Need to apply for Fire Certificate depends on numbers employed, beds provided or materials kept

○ Certificate issued by Fire Authority

○ Certificate must be kept on the premises

A workplace may require a Fire Certificate if it meets certain criteria. An application must be made to the local Fire Authority if the building is used as a factory, office, shop or railway premises in which more than 20 people are working at any one time, or more than 10 persons elsewhere than on the ground floor. In certain cases the Fire Authority now has the power to exempt particular premises from the need for a certificate.

Fire certificates are also usually required for factory premises in which highly flammable or explosive substances are used or stored.

Hotels and boarding houses will require a Fire Certificate if sleeping accommodation is provided for more than six people (employees and/or guests) or any number elsewhere than on the ground or first floors.

Under *Section 10* of the *Fire Precautions Act 1971* the Fire Authority can order the immediate closure of any workplace, or impose severe restrictions on its use, where dangerous conditions are found to exist.

1b: Contents of a Fire Certificate

- Specifies, in respect of premises certified:
  - ○ Means of escape
  - ○ Means for securing the use of means of escape
  - ○ Means for fighting fire
  - ○ Means for giving warning in case of fire
  - ○ Details of highly flammables/explosives (in factories only)
- Much of the information included is in the form of a plan drawing attached to the certificate

The Fire Certificate is prepared by the Fire Authority after an inspection has been carried out and any necessary improvements made. It specifies the fire precautions and procedures agreed at the time the certificate was issued, or subsequently amended. These cover:

- Means of escape in case of fire
- Means for securing the use of means of escape
- Means for fighting fire
- Means for giving warning in the event of fire.

Much of this information is given in the form of a plan drawing, which must be kept with the certificate. The certificate will also state the use of the premises and, in the case of a factory, include details of any highly flammable or explosive materials used or stored.

1c: The Workplace Regulations

- The *Fire Precautions (Workplace) Regulations 1997* (as amended) require employers to provide:
  - Means of escape
  - Adequate means of detecting fire
  - Adequate means of giving warning in the event of fire
  - Fire-fighting equipment
  - Staff training
  - A fire risk assessment

The *Fire Precautions (Workplace) Regulations 1997* – widely referred to as 'the Workplace Regulations' – apply to virtually every place of work, whether it has a Fire Certificate or not. This legislation requires employers to provide adequate:

- Means of escape
- Means of detecting fire
- Means of giving warning in the event of fire
- Fire-fighting equipment
- Staff training

Employers are also obliged to carry out a Fire Risk Assessment, and deal with any problems this uncovers. Staff (or their representatives) should be consulted about fire safety procedures including, where necessary, the appointment of individuals to assume particular roles such as Fire Warden or Roll-Caller. In turn, employees have a legal duty to co-operate with their employer on fire safety matters (this could include drills and the correct use of fire doors) and not to place themselves or other people at risk.

1d: Building Regulations

○   Require plans submission in respect of new building or alterations, including internal partitions

○   Guidance on fire safety standards given in Approved Document 'B'

○   Fire Authority is consulted by Building Control

The design of new buildings, and alterations to existing buildings, are governed by the *Building Regulations*, and these include considerable reference to fire safety, mainly in Approved Document B.

Plans for new buildings, or proposed alterations that will materially affect the means of escape or other aspects of fire safety, must be submitted to the local authority Building Control Department (or an Approved Inspector) and this process involves consultation with the Fire Authority.

In a place of work many aspects of fire safety, such as the specification for fire alarm systems, emergency lighting and fire-fighting equipment, are required to comply with British or European Standards.

Alterations to premises often result in the Fire Risk Assessment becoming out of date, as this needs to be reviewed whenever changes occur in the workplace that could affect the risk of fire. While *Building Regulations* apply at the design stage, ongoing compliance is governed by the *Fire Precautions (Workplace) Regulations 1997* and, where applicable, the *Fire Precautions Act 1971*

1e: Licensing and Registration Arrangements

- ○ Entertainments Licence
- ○ Theatres Licence
- ○ Licence to Serve Alcohol
- ○ Residential Care/Nursing Home Registration
- ○ Nurseries and Childminders
- ○ Many others

Many other pieces of legislation, such as the *Theatres Act, Licensing Act* and *Local Government (Miscellaneous Provisions) Act*, include sections on fire safety provision for activities involving members of the public. These will often be applicable in more general places of work, for instance schools, hotels and community centres.

Premises requiring a licence will usually be subject to inspection by the Fire Brigade or other local authority inspector, who may impose standards higher than those generally applied in the workplace. The Home Office publication 'Guide to Fire Precautions in Places of Entertainment and Like Premises' (available from the Stationery Office) gives useful advice on this subject, but local authorities also publish their own requirements and must be consulted to ensure compliance.

Residential care premises – nursing homes, homes for the elderly, children's homes etc – are generally subject to registration schemes run by the local Health Authority or Social Services Department. Social Services also control Nurseries and Childminders. All of these schemes (and many similar arrangements) involve inspection regimes which include careful scrutiny of the fire precautions.

1f: Change on the Horizon

- A *Regulatory Reform (Fire Safety) Order* due to come into force in 2004:
  ○ Risk assessment based
  ○ 'Responsible Person'
  ○ Consolidating many pieces of legislation
  ○ Dangerous substances
  ○ Fire safety duties

A new *Regulatory Reform (Fire Safety) Order*, under the *Regulatory Reform Act 2001*, is due to come into force in 2004. It will replace the Fire Precautions Act 1971, and the *Fire Precautions (Workplace) Regulations 1997*. Many other pieces of legislation which refer to fire safety will also be repealed or incorporated into this one statute. The need to hold or apply for a Fire Certificate will probably be scrapped.

In this way, fire safety legislation will be reformed to create one simple regime which applies to all workplaces. This regime will be risk assessment based and will identify the person or organisation responsible for the fire safety of the premises. This 'responsible person' (often still the employer) will have to undertake, and act upon, an appropriate Fire Risk Assessment.

As well as employees, the legislation will cover self-employed and voluntary workers, and anyone else on the premises who might be affected by a fire. It is also proposed that employers will have to assess the risks posed by any dangerous substances stored or used, and take steps to control them and mitigate their effects in the event of a fire.

Module 2:
FIRE PRECAUTIONS IN THE WORKPLACE

- ○ Active Fire Precautions
- ○ Passive Fire Precautions
- ○ Fire Safety Management
- ○ The Fire Action Notice

It is important that staff are aware of the fire safety features protecting their workplace, and understand how these are used or what they must be allowed to do.

To understand how a building is protected, it is helpful to think in terms of the 'active' fire precautions (those that actually do something, like the fire alarm system, emergency lighting or sprinklers), the 'passive' ones (like fire-resisting construction and emergency exits), and the management of fire precautions (everything from the setting-up of a company fire safety policy to the routine testing of the fire safety installations throughout the building). The most important thing for staff to know and understand is their Fire Action Notice, a simplified procedure for dealing with fire and evacuation. Familiarity with this notice should form a major part of regular training and drills – another important aspect of fire safety management covered here, and again in more detail in **Module 8**.

2a: Active Fire Precautions (i)

● Fire Alarm and Detection:

○ Call Points

○ Smoke Detectors and Heat Detectors

○ Automatic Door Closers

○ Automatic Call-Out Systems

Identify the fire alarm call points and be sure you would always go to the nearest one if you needed to raise the alarm. Your fire alarm system may also include smoke detectors or − in locations susceptible to steam, cooking fumes or dust − heat detectors. In some premises fire alarm actuation triggers the closing of automatic fire doors, normally held open.

Most modern systems are controlled by a fire alarm indicator panel, which should identify the source of any alarm (or at least the zone in which it was activated) and provide the means for silencing or resetting the system. This should only be done by an authorised person, in accordance with the fire routine for your premises. Some fire alarms are connected to a system that calls out the fire brigade automatically whenever the alarm actuates. If the alarm sounds during working hours, someone should always follow it up with an emergency call to the fire brigade, to confirm the presence of a fire or pass on any available information.

2a: Active Fire Precautions (ii)

○ Emergency Lighting

○ Illuminated Exit Signs

Emergency lighting allows people to escape from the premises in the event of a fire or other emergency, even if this coincides with a failure of the normal lighting system (this is not uncommon, as a large percentage of fires are caused by electrical faults).

The lighting units are designed to come on (or stay on) when the power fails, and to stay on for a period of between one and three hours using an internal battery or other secondary electrical supply. They often take the form of illuminated Fire Exit signs, but can also be seen as additional lighting units on the walls or ceiling of an escape route, or even fully integrated into the normal room or other lighting of a building – a certain percentage of bulbs or fluorescent tubes being provided with this backup. The red glow of a small LED light is often, though not always, an indication of emergency lighting.

Carrying out an effective Fire Risk Assessment will often highlight the need for additional emergency lighting, perhaps more wide-ranging than the standard adopted when the building was first occupied or certified.

2a: Active Fire Precautions (iii)

○ Fire-Fighting Equipment

○ Fire Suppression Systems

Get to know exactly what fire extinguishers and other fire-fighting equipment you have on the premises, and what kinds of fire they can safely be used on. Lift extinguishers off the wall occasionally to get the feel of them, but be prepared for the weight of the larger ones. Extinguishers should be fitted with both a safety pin and an anti-tamper seal – if the seal is found to be broken it must be reported immediately.

Hose reels are provided in some buildings to supplement (or occasionally to take the place of) fire extinguishers. However, injury or damage can easily be caused by incorrect use of hose reels, and common advice nowadays is to leave these for the fire brigade's use only. **Module 7** gives useful information on the correct ways of tackling different types of fire.

As well as portable equipment, there may be sprinklers or other suppression systems (gaseous, powder or water mist) protecting the whole of your premises, or perhaps certain areas such as a computer suite or cooking range. People whose safety, or whose response to an outbreak of fire, depends on the correct operation of these systems must receive the necessary instruction and training.

2b: Passive Fire Precautions (i)

○ Fire Doors

○ Smoke Seals

○ Glazed Door Panels

○ Self-closing Devices

Be clear which doors are fire doors, as these need to be kept shut, unless they are fitted with automatic self-closing devices linked to the fire alarm system. Most fire doors are fitted with a simple self-closing device, and any glazing incorporated will be fire-resisting (eg Georgian wired glass or Pyran). Fire doors should always have smoke seals fitted along their top and side edges.

In most places of work fire doors will be labelled 'Fire Door – Keep Shut'. However fire doors leading to cupboards, storerooms or plant rooms are usually kept locked, and should be labelled with a sign to this effect.

Fire doors are important because they hold back fire and smoke, and are designed to protect means of escape such as corridors and stairways. They also prevent fire spreading rapidly from one area to another.

People should be actively discouraged from wedging or propping open fire doors. Always make sure they are closed when work ceases, and before locking up the premises. Self closers may need adjusting from time to time, if doors are not closing fully into their frames.

2b: Passive Fire Precautions (ii)

○   Fire Exits

○   Locks and Opening Mechanisms

○   Exit Signage

Know the location of all fire exit doors, and how to open them in an emergency. These are the external doors allowing people to escape to open air. Depending on their location, they may be fitted with any one of a variety of opening mechanisms, including panic bolts, push pads, thumb-turns and security devices. However, their operation (in the direction of escape) must not depend on the use of a swipe card, security code or removable key.

Fire exits will usually be clearly marked as such (by a sign incorporating the 'running man' symbol), except for any door that forms the normal means of entering and leaving the building.

2b: Passive Fire Precautions (iii)

○ Fire-resisting Construction

○ Compartmentation

○ Fire-stopping

Many of the structural elements of a building are designed to provide, or benefit from, a particular standard of fire resistance. This will usually be specified at the planning stage, or at the time of subsequent alterations.

To protect means of escape, the normal specification is 30 minutes (for example in walls, doors and glazing forming a protected corridor or stairway) though in some cases this is increased to 60 minutes or even higher, depending on the location and use.

The comparatively safe areas created by this method are known as 'compartments', being separated from all other parts of the building in a way that will limit the spread of fire and smoke throughout the building. This will allow time for the occupants to escape and assist the fire brigade in fighting the fire safely and effectively.

For day-to-day fire safety, the important thing is to ensure that the fire-resisting construction is not neglected, rendered useless by unauthorised alterations, or breached by new pipe work, ducting or other services without being properly 'fire-stopped'.

2c: Fire Safety Management (i)

○ Policy

○ Procedures

To safeguard employees, to protect premises and contents, to ensure continuity of the business and comply with legal obligations, it is essential that a Fire Safety Policy is drawn up and fully implemented.

The policy must detail the agreed procedures for responding to an outbreak of fire or actuation of the fire alarm. It should detail the responsibilities of managers and others with specific responsibilities (such as fire wardens or marshals) and describe the fire safety features of the premises and procedures for regular testing and maintenance. The policy manual should incorporate plan drawings of the premises, detailing the means of escape, assembly point(s) and fire precautions. This could be a copy of the Fire Certificate plan or perhaps one drawn up to accompany the Fire Risk Assessment. Like all aspects of the policy and procedures, this plan must be kept up-to-date. The procedures must be disseminated to all staff at all levels, with a simplified summary of the evacuation procedure encapsulated in the Fire Action Notice.

2c: Fire Safety Management (ii)

- Training
- Testing and Maintenance
- Record Keeping
- Fire Risk Assessment

Regular staff training and evacuation drills are a legal requirement for any business. This should usually be carried out every six months, though in some professions such as healthcare it is normal to train night staff every three months. Different levels of training will of course be required for staff with different levels of responsibility, and at least some staff should receive practical training in the use of fire-fighting equipment.

All training and drills should be recorded in detail, as should testing and maintenance of fire safety equipment and systems. This information should be kept carefully in a Fire Log Book, ideally stored with the Fire Certificate and/or risk assessment, and be readily available for inspection by a local authority fire officer or other authorised person.

The policy must make provision for carrying out, and reviewing, the Fire Risk Assessment as required under the *Fire Precautions (Workplace) Regulations 1997*.

An effective fire safety policy will result in well informed management and staff, avoidance of fire damage and disruption, and a 'culture of fire safety' at all levels of the business.

2d: The Fire Action Notice

○   Simple, clear and concise

○   Action to take on discovering a fire

○   Action to take on hearing the alarm

○   Location of Assembly Point

○   Separate instructions for certain groups/individuals?

Fire Action

If you discover a fire:

- Operate the alarm using the nearest break-glass call point
- Consider fighting the fire, only if you have been trained and it is safe to do so
- Leave the building by the nearest exit and report to the Assembly Point

If you hear the fire alarm...

- Leave the building by the nearest exit and report to the Assembly Point
- Do not stop to collect belongings
- Do not use the lifts

Assembly Point located at...

- Supermarket car park opposite front entrance of this building

You should find copies of the Fire Action Notice strategically placed around the building, preferably adjacent to the fire alarm points (though possibly in other locations too, such as the staff notice board or, in residential premises, on the back of bedroom doors).

The fire routine it describes should be simple and clear. It should say what to do if you hear the fire alarm sounding, or if you discover a fire. Read it from time to time to make sure you understand it, and that you agree with what you are asked to do. Sometimes Fire Action Notices are found to be hopelessly out of date because nobody bothered to check them when changes were made to the building or its surroundings.

The notice should clearly indicate the correct Assembly Point for people evacuating the building – incorporating a floor plan if there is likely to be any difficulty in finding the place.

The Fire Action Notice should form the basis for every evacuation drill: use the drill to test out the procedure, and discuss it with other staff in the de-brief or training session that should follow.

Module 3:
MEANS OF ESCAPE

- Exit Doors
- Protected Routes
- Travel Distance
- Signage

In any workplace it is important that there are sufficient means by which the occupants can make a safe escape in the event of fire. It should be possible for those present to simply turn their back on the fire and escape by their own unaided efforts to a place of safety.

In general terms, means of escape should comply wherever possible with the standards set down in *Building Regulations Approved Document B*, or other guidance issued for the sort of activity carried on, but this may be impossible to achieve in an older building. Your workplace may even be governed by a Fire Certificate issued many years ago when standards were considerably lower than today.

However none of this (not even a valid Fire Certificate) lessens an employer's obligation to carry out an appropriate Fire Risk Assessment and deal with the findings.

3a: Exit Doors

○   Sufficient exits of suitable width and disposition

○   Clearly signposted

○   Easily openable without the use of a removable key

○   Marked on outside 'Fire Exit Keep Clear' if necessary

○   Kept clear of obstructions inside and out

○   Open outwards if more than 50 people may use it

The size and distribution of final exit doors must take into account the number of people occupying the premises, the activities carried on and whether there is a need to provide for wheelchair users. It is important that exits are clearly marked (with the possible exception of the main entrance) and kept clear of obstructions inside and out. Where there is a danger of obstruction occurring outside (eg by parked vehicles) the outer face of the door should be marked with a sign 'Fire Exit – Keep Clear'.

Fire exits must be kept easily openable from within, while anyone is on the premises, and this should not involve the use of a removable key. Where a security device exists on such a door, in must be capable of being over-ridden in an emergency to allow egress without the use of a swipe card or security code.

Where exits are likely to be used by more than 50 people, where they lead from a high risk area or the foot of a stairway, or in places of public assembly, then they should open in the direction of travel.

3b: Protected Routes

- ○ Corridors, lobbies, stairways
- ○ Clear exit width maintained
- ○ No unauthorised storage
- ○ No combustible materials
- ○ Fire doors all working and kept shut
- ○ No breach in fire resisting construction

Where rooms open onto a 'dead-end' corridor (that is, offering only a single direction of escape), or onto a corridor that serves sleeping accommodation, or where there is only one route out of the building, it will usually be necessary to make this a 'protected route'.

This means ensuring that the walls, ceiling and floor all give at least 30 minutes' fire resistance, that there are no breaches in the fire-resisting construction, and that the fire doors are kept closed, unless held open by an automatic system.

It is important that protected routes are kept free of combustible materials and furnishings, or any heat-producing equipment like photocopiers and portable heaters. Notice boards and display stands are easily ignitable – if allowed, any loose papers should be protected by glass or lamination.

In addition a clear exit width must be maintained throughout the escape route at all times, if necessary by marking the floor with yellow hatching. This width is generally defined as 1 metre minimum, or 1.2 metres where wheelchair users may use the escape route.

3c: Travel Distance

○ More than one direction of escape?

○ Type of risk – high, normal, low?

○ Sleeping area?

○ Production area?

○ Residential care?

○ People with disabilities?

When assessing the available means of escape from a building, it is necessary to take into account the distance that occupants will have to walk before reaching a final exit. The distance considered acceptable will vary in accordance with a number of factors, but maximum 'travel distances' have been developed and refined over many years and published in a wide range of government-approved fire safety guides.

The maximum distances shown below are guidelines only, and the risk assessment process should lead you to consider anything else that may have a bearing on someone's ability to leave the premises quickly and safely.

Risk Category	High	Normal	Normal (Sleeping Area)	Normal (Factory Production Area)	Low
Single Direction of Escape:	12m	18m	16m	25m	45m
More than one Escape Route:	25m	45m	32m	45m	60m

3d: Signage (i)

○   'Running man' exit signage (with directional arrows as required)

○   'Fire Exit Keep Clear'

○   'Fire Door Keep Shut'

○   'Keep Locked Shut'

○   'Automatic Fire Door – Keep Clear'

○   'Push to Open'

○   'Break Glass to Open'

The most important signs are those indicating the means of escape, usually consisting of Fire Exit signs (with the 'running man' symbol) on or above designated final exits, and similar signs (with directional arrows if necessary) indicating the route to a final exit. These should be of an appropriate size to be readily seen, and may be internally illuminated as part of the emergency lighting system – this is often a requirement in areas used by members of the public.

Like all the signage discussed in this Module, Fire Exit signs must comply with the *Health and Safety (Safety Signs and Signals) Regulations 1996*. There are a number of commercial suppliers who provide illustrated catalogues, usually free of charge. Studying one of these is an excellent way of finding out about the range of signs in use.

Fire doors should be labelled with a round blue sign at about eye level, on either face, stating 'Fire Door – Keep Shut' or, where applicable, 'Automatic Fire Door – Keep Clear'. Fire doors on cupboards, stores and plant rooms usually require a sign 'Fire Door – Keep Locked Shut', or similar wording.

3d: Signage (ii)

○ 'Fire Action Notice'

○ 'Fire Point'

○ 'Hose Reel'

○ 'Fire Alarm Call Point'

○ 'Fire Assembly Point'

○ 'Disabled Refuge'

Escape routes should not incorporate any obstacle to their use, so instruction signs such as 'Push to Open', 'Break Glass to Open' or 'Press Switch to Open' must be provided where necessary.

Some signage is only required where the layout of the premises makes identification difficult: signs indicating the position of call points, extinguishers (often combined to make a 'Fire Point') or hose reels are examples of this. Signs describing the type and correct use of extinguishers at a particular location are not obligatory, though many employees find these a useful and reassuring 'aide-mémoire'.

However it is advisable to signpost assembly points (as well as describing them in the Fire Action Notice) and disabled refuge areas (see **Module 4**).

This list is by no means exhaustive, and the Fire Risk Assessment may well suggest further opportunities for effective use of signs. Where doubt exists, talk to colleagues and visitors, and decide whether any guidance or clarification is required.

Module 4:
PEOPLE WITH INDIVIDUAL NEEDS

○   It could be YOU!

○   Personal Emergency Egress Plans (PEEPs)

○   Refuges

○   Evacuation Lifts

○   Other Possibilities

There has been for some years a growing trend, supported by powerful and far-reaching legislation, to ensure people with disabilities have full access to places of work, shops, educational and other public buildings. This clearly implies a requirement to include the same people in any planning for escape from fire. Their needs should be reflected in the findings of any Fire Risk Assessment, in the fire precautions installed and the fire and evacuation procedures adopted.

It is important that Fire Wardens (or others with specific responsibilities relating to fire) know about people in their area who may need help in the event of an evacuation, and that this information is passed to the relevant management or security staff who can make special arrangements if necessary. If information isn't volunteered, for example in the case of a visitor with a mobility impairment, it may be necessary to approach the person tactfully to discuss what assistance could be offered in the event of the fire alarm sounding.

People who will be unable to leave unassisted may, by prior arrangement, have a member of staff assigned to them as a helper.

4a: It could be YOU!

○ Mobility impairment – temporary or permanent

○ Loss of hearing or sight

○ Learning disabilities

○ Heart, breathing or other medical conditions

○ Advanced state of pregnancy

There must be arrangements in place for the safety of anyone with a disability or other individual need, either working in or resorting to your place of work.

This will include people with a range of requirements, and not only those who might be considered 'disabled'. Wheelchair users and others with walking difficulties; anyone with a visual or hearing impairment, learning disability, heart or breathing problem; or someone with a temporary condition such as a sprained ankle or advanced state of pregnancy, may need assistance to leave the building.

Whatever the arrangements, make sure that they are understood by the people who will be affected by them, and practised as part of the regular fire drills carried out.

4b: Personal Emergency Egress Plans (PEEPs)

- Consult with employee first
- Discuss nature and location of work, needs, problems and feasible solutions
- Involve line managers, fire wardens, and any potential helpers
- Agree procedures, escape routes, refuges and other relevant details
- Draw up a written PEEP document, including plan drawing if necessary
- Practice agreed procedure during Fire Drills

A Personal Emergency Egress Plan (PEEP) should be drawn up in collaboration with any employee who has a relevant disability or need – temporary or permanent. First, an assessment must be undertaken (by a Health and Safety Manager, Fire Safety Officer or other member of staff with appropriate knowledge and skills), consulting closely with the person in question to determine the level of disability and how this might affect safe evacuation.

Discussion should include the nature of their work and the location(s) where it is carried out, the particular needs and problems of the individual employee and feasible solutions.

Once this assessment has been completed suitable measures should be agreed and written into the PEEP, with copies made available to anyone else who will be involved, such as a line manager, fire warden or designated helper. If necessary a plan drawing should be incorporated into the document showing, for example, viable alternative escape routes.

4c: Refuges

○ Protected landing, lobby or corridor

○ Minimum half hour fire resistance

○ Alternative direction of escape

○ Room for others to pass on way out of building

○ Pre-arranged, agreed and clearly identified

○ Helper stays with disabled colleague?

○ Means of communicating with persons conducting the evacuation

○ Details passed to Fire Brigade on arrival

Refuges are designated areas, protected by fire-resisting construction, in which people with impaired mobility can await assistance. Refuges should, if possible, be sited on each floor adjacent to the stairways forming the escape route. Alternatively, a protected corridor, or even the stairway landing itself, may prove suitable. However, use of an escape route for this purpose must not hinder the evacuation of other occupants.

Each refuge should accommodate the number of disabled people located on that floor, including wheelchair users, and include an alternative direction of escape. It should also have some means of communicating with the Security Desk, Reception or whoever controls the evacuation in your workplace.

A trained team can then help the evacuee to safety by means of an approved evacuation lift or other method. Where delay may occur in carrying out the evacuation, it will be very reassuring to have someone else remain with the evacuee until the situation is resolved.

Refuge areas should be adequately signposted, and clear instructions for use provided within.

4d: Evacuation Lifts

- ○ Must comply with British Standard
- ○ Fire resisting construction
- ○ Backup power supply
- ○ Return to ground or other pre-arranged floor when alarm actuates
- ○ Must be operated by competent person with key
- ○ Do not respond to floor lift-call buttons
- ○ Can be incorporated into PEEPs

Some buildings have one or more lifts that are designated 'evacuation', or sometimes 'fire-fighting' lifts, and these can be used for disabled evacuation if necessary. However, this can only be done by prior arrangement, as it will be necessary for a trained member of staff to take control of and operate the lift. Once the fire alarm sounds the lift will no longer respond automatically to landing call-buttons, and will generally return to the ground floor or other pre-determined level. A key is usually needed to control the lift in these circumstances.

Where such lifts exist it is clearly important to incorporate them into the fire and evacuation procedures for the premises, and the Personal Emergency Egress Plan (PEEP) of any employee likely to require this facility. They should be clearly identified as evacuation lifts, and instructions for their use in an evacuation should be posted inside the car.

Evacuation lifts are designed to meet very stringent safety standards, and have to incorporate fire-resisting construction and a backup power supply. They are far more expensive to install than standard lifts, and for this reason are not often retro-fitted in existing buildings.

4e: Other Possibilities

- ○ Evac-Chair
- ○ Radio operated alarm pager
- ○ Flashing beacon
- ○ Vibrating pillow
- ○ Temporary relocation
- ○ Buddy system

Many other arrangements exist to help people with individual needs escape safely from the workplace, including use of equipment such as the 'Evac-Chair' for helping disabled people down a flight of stairs – remember you must not normally use lifts in the event of a fire.

Fire alarm systems can often incorporate flashing beacons, vibrating pagers (vibrating pillows in sleeping accommodation) or some other system to warn deaf people that the alarm has actuated. Or you may have procedures to ensure that a disabled person will always be accompanied by someone who can lead them safely out of the building.

In the case of someone suffering from a temporary injury or medical condition, it may be possible simply to relocate them to another part of the workplace affording simpler escape routes. However this is less likely to be acceptable in the case of permanent disability.

It must be remembered that where current legislation requires access for people with disabilities, this also entails provision of appropriate means of escape (including alternative means where necessary) equivalent to that specified for any other user of the building.

Module 5:
CAUSES OF FIRE IN THE WORKPLACE

○ Electrical Equipment and Wiring

○ Kitchens

○ Smoking

○ Flammable Liquids

○ Arson

○ Poor Housekeeping

○ Contractors' Work

There are many different causes of fire in the workplace, most of them easily avoidable where there exists an awareness of the dangers, among employers and staff alike, and a determination to keep fire in its place.

Most workplaces are well provided with the means of preventing fire, or limiting the damage an outbreak can do. Yet if fire doors are routinely wedged open, fire drills and training overlooked, false alarms treated as a minor irritation and poor housekeeping allowed in kitchens, workshops and offices, then your business is in serious danger of joining the thousands of others that suffer serious fire losses every year.

On average 500–600 people die in fires every year in this country – though generally very few of these in the workplace – and thousands more are seriously injured. Statistics indicate the main causes to be among the subjects covered in this Module. But there are many other potential causes, and it is the job of the person carrying out the Fire Risk Assessment to look closely, and realistically, at every aspect of the premises, work processes, people at risk and precautions in place. Identify the fire hazards in your workplace, and put controls in place now to avoid becoming another statistic.

5a: Electrical Equipment and Wiring

○ Faulty or untested equipment

○ Damaged wiring

○ Overloaded sockets

○ Unauthorised repairs

○ Inappropriate fuses

○ Wiring run across walkways or under carpets

Faulty or damaged electrical equipment and wiring, is one of the biggest causes of fire in the workplace. All electrical equipment (including items brought in by staff such as radios or kettles) must be properly maintained, and checked regularly in accordance with the *Electricity at Work Regulations*.

Servicing and repairs should only be carried out by a qualified electrician or other competent person.

Sockets must not be overloaded – always fit additional sockets or cable runs in preference to using adapters and extension cables. Blown fuses must only be replaced using fuses or fuse wire of the correct rating.

Never run electric cables under carpet, or anywhere they may be walked on unprotected: normal wear, tear and friction can cause breakdown of the insulation and lead to short circuits and over-heating, often undetected until it results in a fire.

5b: Kitchens

○ Cooking left unattended

○ Care with chip pans and deep fat fryers

○ Fire blanket and appropriate extinguishers available

○ Staff to be properly trained

○ Heat detector preferable to smoke detector

○ Suppression system installed?

Cooking left unattended is a frequent cause of fire. In particular a pan of oil left to heat up for too long can burst into flames and cause serious damage to the kitchen, as well as endangering the occupants. Staff must be made aware of the dangers, and shown safe procedures for using chip pans, deep fat fryers and other potentially hazardous equipment.

Kitchens should also be provided with the correct fire-fighting equipment and staff should be instructed in its use – a chip pan fire can often be tackled safely, but only by someone who has been properly trained.

There may be a suppression system installed (for example high pressure water mist or dry powder) above cooking ranges. This may need to be operated manually, so again make sure its use is properly understood, or contact the suppliers for further instruction.

To avoid false alarms being caused by innocent cooking fumes and steam, heat detectors are usually fitted in kitchens, in preference to smoke detectors. However this will not solve the problem if fire doors are left open allowing fumes to spread beyond the kitchen to the nearest smoke detector.

5c: Smoking

○ Designated smoking area or room

○ Ashtrays and metal bins

○ Regular disposal of waste

○ Harsh policy may encourage illicit smoking

Every employer should have a smoking policy, clearly setting out those areas, if any, where smoking is allowed. In many workplaces smoking is completely banned and smokers have to leave the building to smoke. However if this policy is seen as too harsh, some people may try to get around it by smoking in toilets, storerooms or other quiet corners, where the careless disposal of their cigarette ends could create a more serious fire hazard.

Where smoking is permitted, for example in bars, cafeterias or designated smoking rooms, metal ashtrays or bins should be provided for cigarette ends and spent matches, and these should be emptied frequently into a metal dustbin. Make sure this waste does not come into contact with other refuse until it has had time to cool down.

When carrying out your Fire Risk Assessment do not be tempted to dismiss the hazard with a simple phrase like 'no smoking allowed on premises' – it is important to decide whether any smoking (permitted or otherwise) is likely to occur, and assess the risk realistically.

5d: Flammable Liquids

○ Wide range of everyday substances

○ Must be safely handled and stored

○ Minimum quantities in use

○ Containers clearly labelled

○ Care when decanting

○ Avoid sources of ignition

Flammable liquids give off vapour, and it is this flammable vapour (often spreading far beyond the surface of the liquid itself) which constitutes the main danger. Petrol for example is a highly volatile liquid, while others like diesel or heating oil – though still flammable – are less likely to be accidentally ignited.

Solvents, paints and varnishes are examples of commonly used fluids which may be highly flammable. Look for details on the container and if necessary keep securely stored when not in use.

Try to ensure that the minimum quantities are in use at any time, that containers and stores are clearly labelled with the necessary warnings, and that staff are properly instructed in their safe handling and use.

Do not allow flammable liquids to be used, or decanted, anywhere near a naked flame or other ignition source such as electric motors or engines, and always replace the lid or stopper on the container immediately after use. Remember, the warmer the atmosphere, the greater the spread of flammable vapour.

5e: Arson

- ○ Over 50% of fires deliberately started
- ○ Often associated with vandalism or burglary
- ○ Avoid creating opportunities for fire setting
- ○ Care with waste disposal, wheelie bins, skips etc
- ○ Improve site security and visibility
- ○ Consider CCTV

For most businesses, arson prevention is a matter of good site security, and keeping to a minimum the opportunities for starting a fire maliciously. Obviously not all businesses can afford elaborate security arrangements, but by taking sensible measures to avoid break-ins and vandalism you will also be going a long way towards deterring the would-be fire setter.

Make arrangements for all visitors to report to a reception or security point, and encourage staff to challenge anyone they don't recognise.

If stored outside, keep rubbish and other combustible materials well away from the building, in locked bins or containers if necessary. An overflowing rubbish bin or skip is an obvious temptation to any vandal with a box of matches.

Unless occupied, premises should be locked up securely at night, but remember that a yard surrounded by a solid fence or tall hedge will often mean an ideal location for arsonists to act unseen. Other security systems, such as intruder alarms, security patrols or CCTV should be provided wherever possible.

5f: Poor Housekeeping

- ○ Irregular waste disposal
- ○ Unauthorised storage
- ○ Obstructed escape routes
- ○ Hidden extinguishers, signs and call points
- ○ Untidy, overflowing desks and workbenches
- ○ Abuse of electric wiring/sockets

'Good Housekeeping' is an expression used by fire safety professionals to mean everyday, common sense precautions that will minimise the risk of a serious fire occurring on your premises, and ensure that everyone can get out safely if it does. As the name suggests, it has much to do with tidiness and keeping the workplace in good order.

Allowing rubbish and combustible materials to build up creates a serious risk of a fire breaking out and perhaps spreading rapidly. Arrangements should be made for it to be collected regularly and disposed of safely.

Faulty electrical wiring, or equipment which is abused or badly maintained, presents another risk, as does the inappropriate use or careless storage of flammable materials.

Do not place goods, equipment or furniture where they will hide fire alarm points or extinguishers, or where they will interfere with the operation of fire doors, sprinklers or smoke detectors. Escape routes, such as stairways and corridors, must be kept clear of obstructions or flammable materials at all times.

5g: Contractors' Work

○ Safe Method of Working

○ Hot Work Permit System

○ Control and Supervision

○ Additional Fire Precautions

○ Need to Isolate parts of Fire Alarm System?

Building and maintenance work, particularly any involving 'hot work' such as welding or cutting, is a frequent cause of fire in the workplace. A written Hot Work Permit system should be in place. This involves agreeing a safe method of working prior to commencement, and can include specifying the equipment to be used, the degree of supervision required, additional safeguards to be put in place and so on.

All contractors should be made aware of the fire and evacuation procedures for the premises, and should generally respond to any actuation of the fire alarm as your own staff would.

Measures should be taken to avoid accidental operation of the fire alarm – if necessary by temporarily capping detector heads or isolating sections of the alarm system. It is of course vital to ensure that the system is fully reinstated when work is complete, and between periods of working.

A safety check of the area should be made by staff at the end of each working day, paying particular attention to flammable materials in use, heat-producing equipment and any other potential source of fire.

Module 6:
THE SCIENCE OF FIRE

○ The Triangle of Fire

○ Ways of Extinguishing Fire

○ Making the Right Choice

○ Know your Colour Code

○ How Fire Spreads

This is not a Module dealing with chemical formulas or advanced notions of physics. By the 'science of fire' we simply mean a few very simple analogies that help us break down the mystique of a phenomenon long recognised as Mankind's best friend – and worst enemy. The Triangle of Fire is a simple illustration of the factors that are necessary for a fire to start, and to continue burning. This concept is useful because it reminds us that there are three fundamental approaches to the task of extinguishing fire, each equivalent to removing one side of the triangle. Once established, fire spreads readily from one combustible item to another by direct application of flame. But fire can spread more easily, and often more quickly than this – across a room, around a building or even outdoors – in several different ways. By understanding these processes, it helps us to avoid some of the causes of fire, and take steps to ensure that any outbreak of fire is prevented from spreading unchecked.

6a: The Triangle of Fire

○   For combustion to occur, and continue, three crucial elements must be in place:

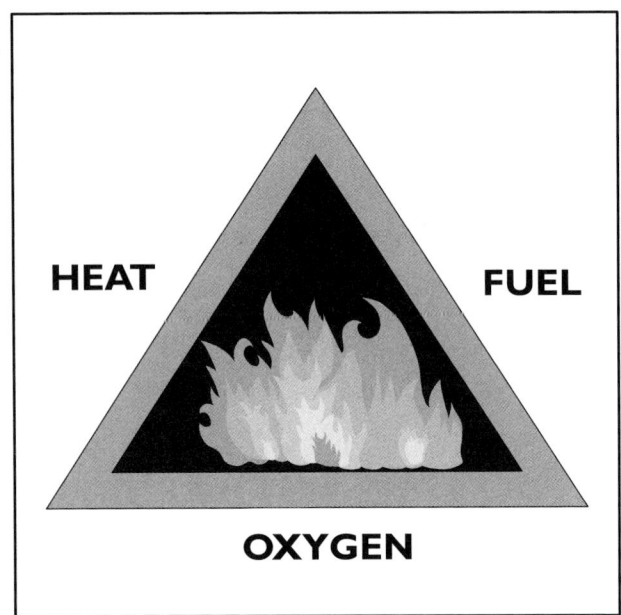

Triangle of fire

There must be a source of **heat** to ignite a fire and this can take many forms: a lighted match, over-heated electrical wiring, a chemical reaction or the sun's rays magnified through broken glass, to name just a few. This heat must also be sustained to allow burning to continue, though this is usually an effect of the combustion process itself.

In this context **fuel** does not just mean petrol, gas, lighter fluid etc, but any material that will burn in response to the application of heat or flames: wood, paper, cardboard, plastics, textiles, fabrics, cooking oil, grass – the list is of course endless, and given the right circumstances even metals and supposedly non-combustible materials will burn.

There must also be a source of **oxygen** to sustain combustion. This is present in the air around us, but can also be created by a chemical reaction in certain materials known as 'oxidising' – perhaps caused by accidental mixing of substances or the careless application of water. Where the oxygen in the atmosphere is enriched, due to oxidising or the use of bottled oxygen (as in some hospital situations), this can cause the fire to burn far more fiercely than would otherwise be expected.

6b: Ways of Extinguishing Fire

- ○ COOLING – reducing temperature
- ○ STARVING – limiting available fuel
- ○ SMOTHERING – depriving of oxygen

Cooling a fire to the point where combustion can no longer be sustained may sound rather complicated, but this is in fact one of the most common ways of putting out fire, eg by applying water to take the heat out of the flames. In some cases ignition can also be prevented by keeping materials cool, for example in the case of flammable liquids, as described in **Module 5**.

A fire will only burn for as long as it has fuel to consume, and removing the available fuel – or **starving** the fire – is a very effective method of fire fighting where the opportunity arises. This might involve moving pallets of readily combustible material away from the path of a spreading fire, cutting a fire-break to limit the spread of a large forest fire, or simply using a rake to pull apart a pile of rubbish ignited by a cigarette end. When there is nothing left to burn, the fire will die out.

Excluding the oxygen, or **smothering** a fire is often the best, and safest, way of putting it out, particularly in the case of burning liquids. Using a fire blanket to extinguish a chip pan fire, and almost any use of dry powder, foam or CO_2 extinguishers, are all examples of fighting fire by smothering the flames.

6c: Making the Right Choice

- Class A (Solids):
○ Water
- Class B (Liquids):
○ Foam, CO_2 or Dry Powder DO NOT use Water
- Class F (Oils/Fats):
○ Wet Chemical ('Liquid Agent'), Fire Blanket DO NOT use Water
- Electrical Fires:
○ CO_2 or Dry Powder DO NOT use Water
- Gas or Metal Fires:
○ LEAVE – Require Specialist Equipment and Training

Having established the methods available for extinguishing a fire, this now helps us to understand the role played by the different types of fire-fighting equipment found in the workplace, and guides us in making the right choice.

It is particularly important to remember that water, while it is the most commonly used fire-fighting medium and the only effective method of cooling a fire, is not suitable for all types of fire. In some cases the use of water could have disastrous, and even fatal, consequences.

Extinguisher labels give instructions on the type(s) of fire they can be used on, and this may include a letter or letters used to classify fires. The most common classifications are shown opposite.

6d: Know your Colour Code

○   WATER: Red Body and RED Label

○   FOAM: Red Body and CREAM Label

○   DRY POWDER: Red Body and BLUE Label

○ CO_2: Red Body and BLACK Label

○   WET CHEMICAL (or 'Liquid Agent'): Red Body and YELLOW Label

Modern extinguishers are painted red, with a small patch of colour visible from the front to identify their contents:

- Red: Water
- Black: Carbon Dioxide (CO_2)
- Cream: Foam
- Blue: Dry Powder
- Yellow: Wet Chemical (or 'Liquid Agent')

Older extinguishers may not comply with this code: for example the entire body may be painted in the colour listed above, or it may be finished in stainless steel. But in all cases the label should state the contents in words, and also remind you – in words or symbols – what type of fire it can be used on.

Remember, no-one should be allowed to fight a fire in the workplace unless they have received adequate, practical training.

6e: How Fire Spreads

○ Convection

○ Conduction

○ Radiation

○ Direct Burning

Hot air rises, and when it can rise no more it will circulate and heat a room, in the process we know as **convection**. Likewise heat, smoke and super-heated gases given off by a fire in one corner of a room will soon spread horizontally across the ceiling. With nowhere further to spread, the hot gases will start to descend, igniting objects far removed from the original source of the fire. If a door has been left open, the fire will push its way out before even reaching this stage, and spread quickly beyond.

The **conduction** of heat, mainly through metal objects such as pipes, ducting, handrails and, if exposed, the structural steelwork of a building, can transfer the heat of a fire from one room, or floor, to another. This can cause fire to break out in locations that would otherwise be unaffected.

By **radiation** we mean heat being transmitted across a space from one object to another. Clothes left to dry in front of a glowing fire will start to scorch, and then ignite, if they are too close. Heat from a fire burning in a warehouse rack may cause damage to other goods stored many metres away. And a large fire consuming one building will often radiate sufficient heat to damage, or set light to, other buildings nearby.

Module 7:
TACKLING A SMALL OUTBREAK OF FIRE

○ The Fire-fighting Equipment in Your Workplace

○ Priorities

○ Fighting Fire Safely

The fire-fighting equipment in your workplace is only designed to extinguish a small fire. Closing the door on a fire, and leaving it for the fire brigade to deal with, will often be a better decision. If you do decide to try tackling the fire, remember the important safety points outlined in the pages of this Module.

Understanding the types of equipment found in your workplace will prove invaluable – knowing, for example, which risks are adequately protected against, whether extinguishers have been properly serviced, or how to identify the extinguisher you need from the colour of its label. But this is no substitute for practical training, and staff who have not received such training should not be encouraged to fight a fire. As a minimum, theoretical training should be backed up with the opportunity to discharge different types of extinguisher (usually water and CO_2) outside in the car park or some other open space. Ideally this should include extinguishing real fires, in a carefully controlled situation. The person carrying out the training should have an appropriate level of knowledge and experience.

7a: The Fire-Fighting Equipment in Your Workplace

- Extinguishers:
- ○ Water
- ○ Foam
- ○ CO_2
- ○ Dry Powder
- ○ Wet Chemical
- Fire Blankets
- Hosereels

Water extinguishers or hose reels should only be used on burning solids, such as wood, cardboard, paper or textiles. **Foam** extinguishers can also be used on this type of fire, but neither water nor foam should ever be used on fires involving live electrical equipment or wiring. For small electrical fires a CO_2 extinguisher is simple and clean to use. A **dry powder** extinguisher is also safe and effective on electrical fires but will leave a great deal more mess to clear up and may damage any sensitive equipment nearby. For burning liquid fires, foam and dry powder are usually the most effective extinguishers, but carbon dioxide can also be used. None of the traditional extinguishers mentioned above are considered suitable for use on a chip pan or other deep-fat fire. A **fire blanket**, or failing this a damp cloth or tea-towel, will often be effective but this should only be used by someone who has been properly trained to tackle this kind of fire. A new kind of extinguisher, generally known as **wet chemical** or **liquid agent**, has been introduced for use on oil and fat fires, and will be found in some commercial kitchens, canteens, restaurants etc.

7b: Priorities

○   Raise the alarm

○   Shout for assistance

○   Assess the type of fire and the dangers it poses

○   Only attempt to fight the fire if you have been properly trained

○   Be sure you have the right equipment for this type of fire

○   Don't tackle large or smoky fires, containers of flammable liquids, spray-cans or cylinders

If you discover a fire your first responsibility is to raise the alarm – usually by operating a call point or shouting to a colleague to do so. Even if your workplace has automatic fire detection installed it may be some time before the smoke or heat reaches the nearest detector head, and people in the building need to be warned at the earliest possible opportunity. If there are visitors or members of the public in the area, explain to them clearly and firmly that they must move away at once. Besides shouting a warning or instructions to others, it will be useful if you can summon a colleague who can assist you in fighting the fire. Meanwhile you should be assessing the fire and the dangers it poses. If it is larger than the average litter-bin fire you should leave it for the fire brigade to deal with. Never try to tackle very smoky fires, or fires involving chemicals, containers of flammable liquids, spray-cans or gas cylinders. Be sure you have the right equipment for the type of fire you are going to fight, and that you have been properly instructed in its use.

7c: Fighting Fire Safely

○ Work with a colleague if possible

○ Test extinguisher before use

○ Crouch down as you approach fire

○ Check your means of escape

○ Use only one type of extinguisher

○ Don't spend more than two or three minutes fighting the fire

Work with another member of staff if possible, either to fight the fire with you or to keep an eye on your safety and your means of escape. Before approaching the fire, test your extinguisher by discharging it momentarily. Crouch down as you approach and begin to tackle the fire, so that heat and smoke can pass harmlessly over your head. You can afford to stand upright when the fire is almost out and you are just damping down the hotspots. Only use one type of extinguisher (for example two water extinguishers or two carbon dioxide). Do not use more than two extinguishers in total, or spend more than two or three minutes fighting the fire with a hose reel. And remember to keep checking that your means of escape is not becoming cut off by spreading fire or smoke.

When using a fire blanket, fold back the corners to protect your hands and wrists. If you are tackling a chip pan or other cooking fire, first turn off the gas or electricity, if you can do so safely.

Workplace Fire Safety

Specific Training Modules 8–12

Every place of work, and every workforce, has to be suitably protected from the dangers of fire. As described in earlier Modules, there must be adequate means of escape (protected where necessary, in accordance with modern principles of fire safety), means of detecting fire and raising the alarm, fire-fighting equipment, emergency lighting, management procedures and training. In all but the smallest workplaces, 'adequate' is likely to be interpreted as complying with British Standards or their European equivalent, or forming part of an engineered fire safety solution drawn up by fire safety professionals and backed up with mathematical and other data. **Module 2, and Part 1** of this manual, described the main points to be observed when drawing up the fire policy for any business. While the contents of the policy will inevitably differ from business to business and premises to premises, the basic principles remain the same. Remember, modern fire safety legislation is based on the requirement to carry out a realistic Fire Risk Assessment, and it is this above all else that should determine the policies, procedures and physical fire precautions needed to safeguard your particular business.

Module 8:
FIRE SAFETY IN COMMERCIAL PREMISES

- Policy and Procedures
- Fire Precautions
- 'It Could Never Happen Here'
- Training and Drills

Although the risk of a serious fire may not seem as great in an office, shop or similar commercial setting, the requirement to provide and maintain fire precautions (whether or not a Fire Certificate exists) is as rigorous and binding as would be the case in a factory, warehouse or other industrial building. Frequently an office block or other largely administrative building will have the advantage of a fairly simple and repetitive layout, generally low-risk work activities, a clear management structure and static workforce – by comparison with a large industrial site, for example, with its complex layout, high-risk production processes, fluid workforce, shift working and so on. It follows that it should be a comparatively simple matter to establish the aims of the fire safety policy – largely in terms of life safety, protection of data, and business continuity – without the complications of production schedules, raw materials storage or monitoring of high-risk activities.

8a: Policy and Procedures in Commercial Premises

○ Establish aims, hierarchy, responsibilities, and list of 'competent persons'

○ Policy to include Fire Risk Assessment

○ Draw up procedures in consultation with workforce

○ Clear, concise, Fire Action Notice to summarise fire and evacuation procedures

Defining a fire safety 'hierarchy' and detailed responsibilities from, say, the Fire Safety Officer or Facilities Manager, down through Section Heads, Security Team, Floor Fire Marshals and individual workers, is a relatively straightforward task in an office or shop environment, though of course one which should involve consultation with the workforce or their representatives. Procedures established should be detailed in the company Fire Manual, disseminated to the individuals or groups concerned, and practised in fire drills carried out on a regular basis.

Fire Action Notices, which must be prominently displayed throughout the building, should list basic instructions on responding to the fire alarm or discovering a fire. In a retail setting, safe and orderly evacuation of the public will be a high priority.

Identification of 'competent persons' (from within the organisation or, if necessary, using outside contractors or consultants) will be necessary for dealing with Fire Risk Assessment, testing and maintenance of fire safety equipment and systems, staff training, and liaison with the fire brigade and other relevant authorities.

8b: Fire Precautions in Commercial Premises

- In-house testing of systems and installations
- Specialist maintenance contracts
- Accurate record keeping (including staff training) in Fire Log Book
- Visual checks and early attention to defects reported
- Fire Certificate and/or Fire Risk Assessment kept up-to-date

All fire safety systems and installations have to be tested regularly, in accordance with the relevant British Standard and/or manufacturer's instructions, and this will usually involve a combination of local testing (for example the weekly test of the fire alarm) and periodic maintenance undertaken by a competent contractor.

Visual checks of fire precautions, such as fire doors, exits and extinguishers, should also be carried out regularly, and all staff should be encouraged to report any defects they notice as they go about their work.

Defects discovered must be attended to as a matter of urgency.

All testing, and the results of those tests, should be recorded in a Fire Log Book, kept available for inspection by an authorised person. Training and drill records should also be kept in this book.

Wherever changes have occurred – to the building, its use, occupancy etc – it is important to ensure that the Fire Risk Assessment is reviewed and if necessary amended. If a Fire Certificate is in force, the Fire Authority must be notified of any material changes, particularly those affecting means of escape.

8c: 'It Could Never Happen Here'

- ○ Fires occur in new buildings as well as old ones
- ○ Quantities of paper, packaging and other combustibles
- ○ Electrical equipment and wiring
- ○ Possible apathy towards fire precautions and drills
- ○ Importance of 'housekeeping'
- ○ Develop 'culture of fire safety'

The comparatively low-risk work activities carried on in a commercial environment often lead to an attitude of complacency among some staff, who find it hard to envisage the consequences of a serious fire, and even see fire safety as so much 'red tape'. The phrase 'It could never happen here' crops up again and again.

Unfortunately, fires occur in new buildings as well as old ones, and in shop or office buildings as well as factories, schools and hotels. In fact some commercial premises have a relatively high 'fire loading' (the quantity of material available to burn and sustain a serious fire) in terms of stock, packaging, stationery, waste paper and filing systems, as well as heat-producing copying and IT equipment.

While none of this necessarily makes the work process 'high-risk', it does emphasise the importance of the fire safety provision, of safeguarding the means of escape, and taking part in fire drills. With a sound fire safety policy supported and put into practice by all levels of management, and with effective, relevant training involving all members of the organisation, then apathy can in time be replaced with a 'culture of fire safety'.

8d: Training and Drills in Commercial Premises

- ○ Fire Safety Management Seminar
- ○ Fire Warden Training
- ○ Specialist Duties (eg Security, Catering staff)
- ○ Induction Training
- ○ Fire Awareness Courses
- ○ Fire Fighting and Extinguisher Handling
- ○ Realistic Evacuation Drills

Training must involve every member of the workforce, and should be appropriate to the level of responsibility undertaken. Clearly a Managing Director, Health and Safety Officer or Facilities Manager is likely to have considerable responsibility for fire safety matters (in legal as well as practical terms) and this should be reflected in the training undertaken: a *Fire Safety Management Seminar* to update knowledge and skills, and reinforce (or review) company policy would be suitable at this level.

Between the two extremes of individual responsibility there may be a range of specialist roles (*Fire Wardens, Security Officers, Catering Staff* etc) requiring individual or group instruction. Personnel nominated for *Fire-Fighting and Extinguisher Handling* (on a voluntary basis generally) should receive both theoretical and practical instruction.

As a minimum, all staff should receive basic *Fire Awareness Training* and a realistic Evacuation Drill approximately every six months, and of course all new staff must be given simple, basic instruction in the fire and evacuation procedures as part of their *Induction Training* on joining the organisation.

Module 9:
FIRE SAFETY IN INDUSTRIAL PREMISES

○   Process Risks

○   Flammable and Explosive Materials

○   Fire Protection Systems

It is commonly assumed that the most serious fires are likely to occur in industrial premises, particularly those used to process, or store, large quantities of flammable or explosive materials. However, if properly managed and adequately protected, then factories, warehouses and other industrial facilities – large or small – should be in no greater danger of suffering loss from fire than an office, hotel or retail complex.

Wherever high-risk materials or activities are found, then appropriate compensating measures should have been included in the fire precautions and procedures built around them. These measures must of course include adequate means of escape, but in fact escape from an open-plan factory or warehouse building is often far easier than from a large office building, with its convoluted corridors, inner offices and phased evacuation procedure.

Again, risk assessment is the key, but there may also be greater need to comply with statutory controls, such as fire certification and petroleum licensing, as well as Health and Safety Executive guidance on the storage and use of certain materials.

9a: Process Risks in Industrial Premises

○ Heat producing equipment or manufacturing process

○ Highly flammable/explosive substances in use

○ Combustible products or packaging

○ Specialist fire-fighting procedures or equipment

○ Fire Risk Assessment may suggest additional precautions

The greatest likelihood of a fire occurring is in the *process* of manufacturing, adapting, treating or packaging goods – particularly where that process involves the use or production of heat, or where the goods, components or materials used are of a highly flammable or explosive nature. Procedures adopted must comply with any instructions provided by the manufacturers of materials and equipment involved in the process, as well as regulations and guidance laid down by the Health and Safety Executive, the Fire Authority (for example in the conditions of the Fire Certificate) and any other relevant statutory or regulatory body.

There may be specialist equipment in place to deal with any outbreak of fire, such as graphite powder extinguishers (for metal fires) or manually-operated suppression systems. In some industries there may even be a trained fire-fighting team available on-site, possibly equipped with breathing apparatus.

All this needs to be reviewed regularly as part of the Fire Risk Assessment monitoring procedure, particularly in terms of adequacy and reliability of protection systems, availability of trained staff, technological progress and current industry guidance.

9b: Flammable and Explosive Materials

○ As specified in Fire Certificate?

○ Stored and handled in accordance with HSE guidance?

○ Included in Fire Risk Assessment?

○ Staff fully aware of risks and safe procedures?

○ Appropriate fire-fighting equipment or installations?

○ Petroleum Licence required?

There are strict guidelines governing the use, handling and storage of highly flammable and explosive materials, many of these having more to do with Health and Safety legislation (for example COSHH, and labelling regulations).

Crucial though this guidance is, it is also necessary to comply with any fire-related regulation, such as the need to obtain a Fire Certificate or Petroleum Licence, and to carry out a thorough Fire Risk Assessment. All of these, where applicable, must be kept up to date: where flammable or explosive materials are found to vary substantially from the types or quantities listed or provided for, the employer (or other 'Responsible Person') could face prosecution.

Necessary provision for the storage of such materials (whether or not stipulated by the relevant authority) might include a properly-constructed flammable materials store, specialist fire-fighting equipment or suppression systems, additional fire detection equipment and, in all cases, proper staff training incorporating any specialist knowledge, skills or equipment required.

9c: Fire Protection Systems for Industrial Premises

○ Sprinklers and drenchers

○ Gaseous suppression systems

○ Water mist

○ Alarm and detection systems

○ Smoke extraction and ventilation

There is a surprisingly wide range of systems that can be installed to protect a workplace, in addition to fire alarm and detection equipment.

Sprinklers give considerable protection, particularly against loss of property, and are often found in manufacturing and storage facilities. Where appropriate, drenchers may be fitted (usually on the outside of a building) – these spray water from every head simultaneously, whereas sprinkler heads usually operate individually, as affected by the fire.

Gaseous suppression systems are frequently used to protect valuable equipment (such as computer suites) or high-risk manufacturing processes. They are likely to employ Halon replacement gases such as Inergen or FM200, or they may simply use CO_2 – extinguishing a fire by flooding the area, within a controlled and generally sealed environment. High-pressure water mist (or fog) systems can also be very effective, and used safely even for electrical and deep-fat frying equipment.

Various smoke extraction and ventilation systems are also commonplace, often greatly reducing the damage a fire may cause, and making fire-fighting operations easier and more focused.

Module 10:
FIRE SAFETY IN HEALTHCARE PREMISES

○ Hospitals

○ Residential Care

○ Day Centres

○ Responding to the Fire Alarm

○ 'Horizontal Evacuation'

Fire safety is crucial to every type of healthcare premises, in terms not only of the importance of such facilities to the community but, frequently, the presence of people who are at greater risk from an outbreak of fire – due to ill-health, mobility impairment, inability to respond to the fire alarm, or the simple fact that they may be asleep at the time.

Of course the term 'healthcare' covers an enormously wide range of premises, often including office and industrial workplaces dealt with already in earlier Modules. However it can safely be assumed that advice given here, in conjunction with the more 'generic' **Modules 1–7**, will cover the most important issues relating to fire safety in hospitals, residential care homes and day-care facilities.

More specific guidance is provided in the suite of documents known as 'Firecode', a range of Health Technical Memoranda (HTMs) and Fire Practice Notes (FPNs) published by NHS Estates. The guidance given in these documents is very comprehensive, and widely applied to both NHS and other healthcare premises, covering a huge range of topics from risk assessment to arson prevention, from kitchens to laboratories, from hospitals to patient hotels.

10a: Hospitals

○ New hospitals meet very high standards of fire safety and compartmentation

○ Escape based on progressive horizontal evacuation

○ Older hospitals may not meet current standards

○ Need to carry out Fire Risk Assessment

○ High-risk areas: pathology laboratories, laundries, kitchens etc

New hospitals are designed to a very high standard of fire safety, including rigorous 'compart-mentation', sophisticated fire alarm and detection systems (each zone generally forming a 'sub-compartment'), and extensive application of 'progressive horizontal evacuation' as the principle governing any necessary evacuation of patients in the event of a fire. Procedures and training take into account the particular needs of bed-ridden or non-ambulant patients, and the probable need to evacuate many patients with their beds.

Older hospitals are often adapted and equipped to comply with modern standards but this is not always possible due to physical and financial restraints. High standards of fire safety provision, housekeeping and training remain paramount, and of course there is no exemption from the need to carry out, and act upon, a full Fire Risk Assessment that takes into account all users of the building.

Particular attention must be paid to high-risk areas, including pathology laboratories, linen stores, laundries, kitchens, boiler rooms and oxygen-enriched atmospheres.

10b: **Residential Care**

○ Includes homes for the elderly, nursing homes, children's homes etc

○ Registration schemes in force

○ High-risk activities to be carefully assessed and controlled

○ Horizontal evacuation and special procedures for responding to fire alarm

'Residential care premises' can include homes for elderly people, nursing and convalescent homes, children's homes, group homes and a range of similar facilities (other than hospitals) providing sleeping accommodation with a high degree of supervision and care, in respect of people with medical, physical, or behavioural needs. They will inevitably be governed by a registration scheme, which usually sets the fire safety standards to be complied with, and may involve a regular inspection regime.

As with hospitals, fire safety design is usually based on a high standard of fire alarm and detection, compartmentation throughout, and adoption of the principle of progressive horizontal evacuation

Smoking, where permitted, must be strictly controlled, but may include arrangements such as designated smoking areas or supervised smoking in bedrooms – these and other high-risk activities (kitchens, laundries etc) must be carefully and realistically risk-assessed.

Fire alarm and evacuation procedures generally vary from those in other places of work, due to the needs of the residents, and may for example include a staff assembly point *inside* the premises.

10c: Day Centres

- A range of client groups, including the elderly, young children and people with disabilities
- Varied activities, may include kitchens, craft studios and hairdressing salons
- Fire safety provision less stringent than residential care
- Full or partial evacuation procedures – must be practised and understood by all

Day centres are usually run by Social Services, the NHS or voluntary organisations, and cater for the needs of a wide range of clients including the elderly, people with physical and learning disabilities, and parents with young children. They vary greatly in size and layout, and may be purpose-built or older, converted buildings – sometimes on several floors.

Activities can be very varied, encompassing art and crafts, music, fitness, bingo and computer use. The use of hairdressing salons, clay kilns and certain other craft facilities may present a higher risk of fire, as will a kitchen providing hot meals.

Having no sleeping accommodation there may be less fire safety provision than in residential care, (eg little or no smoke detection) but again this must comply with the findings of a full Fire Risk Assessment.

In the event of fire alarm actuation, day centres are more likely to be evacuated fully, but the procedures may vary depending on the nature of the group attending. Any partial or delayed evacuation must take into account the compartmentation of the building and availability of refuges.

10d: Responding to Fire Alarms in Healthcare Premises

○ Residents usually remain on the premises

○ Staff assembly point may be inside the building

○ Trained staff sent to investigate

○ Horizontal evacuation may be required

○ Importance of calling fire brigade

○ Avoid false alarms caused by toasters, smoking etc

Due to the unusual nature of healthcare premises, and the vulnerability of their clients, fire alarm and evacuation procedures frequently differ considerably from what would be acceptable in other workplaces. Residents or patients are often encouraged to remain in their bedrooms or other safe areas (such as a lounge) protected by fire resisting construction, while the cause of the alarm is investigated and dealt with.

On hearing the alarm, staff may be required to report to an assembly point *inside* the building (usually by the fire alarm panel) and await further instructions. The person in charge will take the information from the panel and may send trained staff to the affected area to check the situation (but without putting them at risk). Any clients found to be in immediate danger should be moved to safety by staff using any means available, based on the system of 'horizontal evacuation' described below (see **10e** below)

It is important that the fire brigade is called immediately, before investigating, unless there is absolute certainty that it is a false alarm. Activities that can lead to unwanted alarms, such as smoking and the use of toasters, must be carefully controlled to avoid disruption and complacency.

10e: 'Horizontal Evacuation'

○ Avoids need to evacuate premises fully or negotiate stairs

○ Makes use of building compartmentation

○ Most clients left in rooms

○ Clients in danger moved to another compartment on same floor, and again progressively if required

○ Importance of fire doors

Full evacuation of hospitals, or care homes for the elderly or infirm, is usually neither practical nor desirable. At night for instance, with a reduced number of staff on duty, it would be physically impossible to take all residents outside, and probably more dangerous than keeping them in.

Residential care premises are provided with a high standard of fire alarm system, including smoke or heat detection throughout. They are also divided up into protected areas, or 'compartments', to stop fire spreading through the building and create safe areas for people to congregate.

If a fire is discovered, most clients will be safer left in their rooms, or in a lounge or dining area, provided of course that they are protected by a closed fire door. Where necessary, a small number of residents can be moved safely from the affected area into another compartment on the same floor, while the fire is dealt with or until the fire brigade arrives.

This is called 'horizontal evacuation' and will always involve passing through at least one set of fire doors. Further 'progressive' evacuation can take place if the fire situation worsens. There should not usually be any need to negotiate stairs.

Module 11:
FIRE SAFETY IN EDUCATIONAL PREMISES

- ○ Day Schools
- ○ Boarding Schools
- ○ Colleges
- ○ Nurseries

Educational premises are naturally expected to provide and maintain the highest standards of fire safety – particularly those with responsibility for large numbers of children or young people, who may be considered more vulnerable due to unpredictable behaviour and lack of experience. In fact schools and colleges have a very good record of safeguarding their students and staff, applying discipline, leadership and regular practice (eg termly fire drills) to compensate for youthful unpredictability. However, in terms of property, annual fire losses in educational premises (often out of school hours) are extremely high.

Many communities suffer the devastating loss of a local school or other educational building – with subsequent loss and disruption to children, families, staff and other users. In the case of schools, it is believed that some 60% of these fire losses are caused by deliberate ignition – perhaps a malicious arson attack, more often just mindless vandalism. The way to improve this situation is a combination of greater site security, improved fire alarm and detection, regular staff training and, where the need exists, community education campaigns.

11a: Day Schools and Colleges

○ Identify high fire risk activities

○ Importance of staff training and evacuation drills

○ Large quantities of combustible material, including craftwork, displays, notice boards etc

○ False alarms

When occupied, school and college premises are most at risk from illicit smoking, work processes (particularly in science laboratories, craft studios and workshops), kitchens, faulty electrical equipment or wiring, and deliberate ignition – though this is an even greater risk out of school hours, where a maliciously started fire may develop undetected.

It is important that staff are well trained, conscientious and vigilant, and that evacuation drills are carried out (at least termly) and taken seriously.

By their very nature schools and colleges are full of easily combustible materials: mountains of loose paper, books and stationery; displays of art and craft work; posters and written work pinned to classroom walls; and notice boards crammed full of announcements. It is important that this tendency does not encroach onto stairways and protected corridors where these items may represent a combustible hazard or obstruction.

In some establishments false alarms are a regular nuisance – this is a difficult management problem. One solution is the fitting of alarmed, transparent covers over call-points – a loud wailing sound attracts attention to the user and discourages improper use.

11b: Boarding Schools

- ○ DfES inspection regime

- ○ Subject to the *Fire Precautions (Workplace) Regulations 1997* and other legislation.

- ○ Extensive catering facilities

- ○ Sleeping risk requires high standard of fire safety, including comprehensive alarm and detection system

- ○ Include night-time drill, using appropriate assembly area.

In many respects boarding schools face the same problems as day schools, but are required to meet even higher standards of fire safety in view of the sleeping risk present. For this reason they are usually subject to a regular inspection regime, arranged by the Department for Education and Skills. They are also governed by the *Fire Precautions (Workplace) Regulations 1997* (and therefore required to carry out a Fire Risk Assessment) as well as any other applicable fire safety legislation – if, for example they need an Entertainment or Theatre Licence for certain activities.

Catering is likely to be more extensive than in a day school, and large, busy kitchens will require particular attention in terms of fire risk management, compartmentation, and staff training.

Dormitories and other sleeping accommodation should be protected throughout by a well maintained fire alarm and detection system, and by the provision of good means of escape, emergency lighting, fire-fighting equipment and signage, all complying with current standards. Again fire drills should be a regular feature of school life, including at least one drill per year carried out at night. This may highlight the need to arrange an appropriate, sheltered, assembly area.

11c: Nurseries

○   Evacuation procedures must meet the needs of very young children

○   Children to be kept away from cookers, heaters, sockets etc

○   Require good means of escape and other fire safety provision

○   Registered childminders subject to similar requirements

Nurseries are usually registered with Social Services and subject to an inspection regime, possibly carried out by the local fire brigade.

As with all educational premises there must be good means of escape (protected where necessary) and other fire safety provision. But the fire and evacuation procedures adopted must take into account the needs and abilities of very young children, possibly even babies. For this reason nurseries generally require a very high staff-to-child ratio, and careful planning to ensure the safety and security of children during an evacuation. The procedures drawn up must be well understood by all staff and practised regularly, involving real children in any drill carried out.

Premises used as a nursery must have safeguards in place to keep young children away from cookers, heaters, electrical sockets etc.

Arrangements for registered childminding in private dwellings are similar, if somewhat less elaborate. Particular attention must be given to the provision of adequate smoke detectors and means of escape, including self-closing fire doors where necessary to protect escape routes.

Module 12:
FIRE SAFETY IN RESIDENTIAL PREMISES

○ Hotels

○ Hostels

○ Flats

○ Sheltered Accommodation

Apart from healthcare or education premises, residential buildings tend to fall outside of the area normally covered by fire safety legislation. Of course the design of new dwellings is governed by *Building Regulations* and the *Housing Act* allows for considerable control over the letting out of a building for bedsits. But once occupied, a private individual dwelling is beyond the scope of most forms of fire regulation.

This is largely because it does not fall into the category of 'workplace'. Unfortunately fire does not seem to recognise such distinctions, and most of the deaths and serious injuries caused by fire in this country do occur in the home.

At the other end of the scale hotels and boarding houses were the first category of business to have been designated under the *Fire Precautions Act 1971* – and are still required to hold, or apply for, a Fire Certificate if they provide beds for more than six persons (including staff), or any number elsewhere than on the ground and first floors. They are also clearly a place of work, protected by the requirements of the *Fire Precautions (Workplace) Regulations 1997*.

12a: Hotels

- ○ Most hotels require a Fire Certificate
- ○ High standard of fire alarm and detection
- ○ Risks may include residential, catering, retail, entertainment and other activities
- ○ Training and drills essential, and should involve all types of staff – and guests

As already described, hotels generally have to apply for a Fire Certificate, though very small establishments may be governed by the provisions of the *Housing Act*, and inspected for fire safety by the local Environmental Health Officer.

Hotels are required to have a very high standard of fire alarm and detection system throughout the premises, and to operate rigorous fire and evacuation procedures covering staff, guests, diners and anyone else resorting to the premises. While the causes of fire in hotels vary little from other types of workplace, the presence of large numbers of the public makes it more difficult to control activities which may lead to a fire, or the response of occupants to an outbreak or warning of fire.

Staff training is vitally important — covering all areas of hotel activity including catering, retail, security, management and domestic staff — and regular drills must take place, preferably with the involvement of guests. All rooms must be provided with a clear Fire Action Notice, but hotel guests are notoriously complacent in responding to fire alarms and many hotels have a policy of knocking on room doors to clear the building in an emergency.

12b: Hostels

○ Only open to certain groups or categories of occupant

○ Fire safety standards similar to hotels

○ Risks related to nature of occupancy

○ Smoking, cooking and clothes drying among causes of fire

○ Poor housekeeping and lack of supervision may worsen effects of fire

Hostels are often very similar to hotels in the way they are run and the risks they present. But they are not defined as hotels, or issued with Fire Certificates, as they are only open to certain groups or categories of occupant. Examples are student halls of residence, shelters provided by local authorities or charities for the homeless, and nurses' quarters or similar accommodation provided by employers for their staff.

Again a high standard of fire safety is required (including automatic fire detection throughout) and residents must be fully briefed on the fire and evacuation procedures, the importance of fire doors, smoking policy etc.

The nature of the occupancy may well suggest other areas of concern: alcohol or drug abuse, the presence of young children, people with disabilities, or the inability to read or understand instructions given in English.

Typical causes of fire in hostels include late night cooking, carelessness with cigarettes, and unsafe arrangements for drying clothes – while the effects of such an outbreak may be exacerbated by unauthorised storage in corridors, difficulty in rousing sleepers, and the absence of supervisory staff at certain times of the day or night.

12c: Flats

○ Workplace legislation unlikely to apply

○ Fire safety may be to *Building Regulations*, British Standard or other guidelines

○ Designed to let residents stay in their flats, in the event of fire

○ Fire alarm and detection system not usually designed to evacuate block

Flats are not generally classified as a place of work, and again no fire safety legislation is usually enforceable within a single private dwelling. Building Regulations impose many stipulations on the fire safety design of a new building, but older blocks of flats may not comply with these standards. Instead, landlords, managing agents or others charged with the safe management of the building will often address fire safety in accordance with local authority guidelines or British Standards.

Most blocks of flats are designed around protected lobbies, corridors and stairways (sometimes a single stairway). The idea is that it should not be necessary to evacuate the block, as residents will remain safe within their own accommodation, protected by fire resisting construction and fire doors. This emphasises the crucial importance of good maintenance: fire resistance can be eroded by wear and tear, neglect and vandalism.

Any fire alarm system installed is therefore unlikely to evacuate the whole building. It will probably consist of smoke detectors in individual flats (possibly heat detectors where kitchens are not separated) and some coverage of the common areas. All this may be monitored by a warden service, or a system that automatically calls the fire brigade.

12d: Sheltered Accommodation

○ Warden controlled flats or small houses for independent, older people

○ Run more like a block of flats than a care home, including fire alarm and evacuation policy

○ Communal facilities often run on day centre lines

○ Evacuation lifts, if provided, will depend on staff for use in the event of fire alarm sounding

Sheltered accommodation usually comprises a development of flats or small houses, provided with a warden service and designed to allow older people to live independent lives within a community setting. The accommodation may, however, be part of a larger complex shared with a residential care home or other healthcare facility.

Fire safety in sheltered accommodation is governed *by Building Regulations* and British Standard guidance, but again it must be remembered than individual flats or bungalows constitute single private dwellings. Such developments are therefore run more like a managed block of flats than a care home, and this includes fire and evacuation policy. Fire alarm activation within a dwelling will usually alert the warden service, and the fire brigade will be called if necessary, but without evacuating the block or complex. Communal lounges, club rooms or dining areas are likely to be run on lines very similar to a day centre.

Sheltered flats may be provided with evacuation lifts that can be used in the event of fire, but in such circumstances these will only work under the control of a member of staff, usually equipped with a key.

Part 3: Supporting Information

Guidance for the Trainer

Introduction

The use of **Part 2** of this book presupposes that fire safety training is being undertaken in-house. For the purposes of discussion it is assumed that the person who will carry out fire safety training has little or no formal education in fire precautions and that this publication will form the basis of information passed on to staff.

It is strongly recommended that fire safety trainers study the training material fully, in advance, in order to assess and satisfy themselves with regard to their ability and competence to deliver the fire safety training programme.

Several useful sources of information are referred to in **Part 2**, and the potential reading list for those requiring more in-depth study is enormous. Numerous guidance documents dealing with the subject of fire safety in the workplace are published by Government Departments, the Health and Safety Executive, the British Standards Institute and other organisations, and there are also professional journals and magazines. The author advises that this manual alone may not provide sufficient information (as a stand-alone reference source) to deliver adequate training for *all* businesses.

Forms of Presentation

The assumption is made that a traditional approach to fire safety training will be used. This includes training sessions at which members of staff attend and a trainer delivers the message. There are alternative methods that have been developed involving IT systems: use of the desktop PC and an interactive training programme may be involved. On its own this approach may be considered a rather remote method, and aspects of 'culture' and the buy-in of staff may be lost. On the other hand such methods may be a bonus if they form only part of the training regime – perhaps posted on the organisation's intranet, supplemented by copies of safety policy and procedural documen-

tation. In the author's experience few organisations have *all* staff working with or having access to a PC and associated systems.

Venue for Training

It is important in the delivery of training that the setting is right. It is strongly recommended that the room is of sufficient size for the group to be trained, that seats are comfortable and that table or desk space is provided to encourage delegates to make notes and use any pre-prepared materials handed out. Ideally the space should not be the participant's normal place of work (to avoid routine distractions) and if the room has a telephone it should be silenced by prior arrangement. Any other form of distraction or interruption should be strongly discouraged (including mobile phones).

Training Setting

It is important that the trainer makes clear to delegates the status and relevance of the training course, but on the other hand it is essential to create an atmosphere of participation and involvement. Establishing the balance between legal duty, life preservation and general interest is a skill in itself. 'Setting the mood' can be supported in ways that send a signal to delegates, such as how seriously the organisation takes fire safety issues. This can be demonstrated by the presentation of the space, the seniority of delegates, the quality of supporting materials and the allocation of resources – in particular the time of all involved.

Training Duration

For the vast majority of delegates fire safety training is a supplementary activity to their main employment function. With people working under pressure, it is important to keep fire safety training focused and concise. It is recommended that the training courses included within this book be delivered as follows:

- General Training Modules 1–7 (90 minutes)
- Specific Training Modules 8–12 (30 minutes (additional))
- Fire-fighting equipment, practical use (20 minutes (additional)).

If possible, either during or following a training course, arrange refreshments and an opportunity for the delegates to discuss the issues raised, among themselves and with the trainer. Encourage free discussion and lead it where you are able. Offer to provide additional information or support where it is difficult to respond to issues immediately.

Training Materials

It is important that delegates take something away with them on completion of the training session. This can be in the form of copy presentation notes, their own supplementary notes, or other materials considered appropriate to the workplace risk. Delegate packs should therefore be prepared, and this underpins the issue raised above concerning the seriousness accorded by the organisation's senior management.

The presentation may be made using computer generated and projected images, by overhead projector, or purely using handouts. Again it must be stressed that the more trouble that is taken to deliver the fire safety message, the more the delegates will value it. Other supplementary material may be used to expand upon or emphasise areas of particular relevance to your workforce. There are many excellent videos covering much of the material presented in **Part 2**, but over-reliance on 'off-the-shelf' videos is not recommended as this devalues the specific nature of training, and on subsequent 'refresher' courses these quickly become tedious. If they are used it must be recognised that additional time will be required.

Numbers Trained

Several issues will determine the numbers to be trained in each session: the size of the venue, the availability of staff, the ability (or style) of the trainer, the intention to include practical extinguisher training etc. In general, it is recommended that the training group size does not exceed 25. This number allows for interaction with the trainer and enables the trainer to gain some involvement from even the shyest delegates. It is vital that this level of participation is sought and that all are encouraged to 'get involved'. Numbers in excess of 25 are difficult to monitor and control, and may appear quite daunting to someone who is still new to the training role. Public speaking is not a skill people are born with, it has to be learned. If you are apprehensive about delivering training, aim to work with small groups (say 8–12) until you have built up your knowledge and confidence.

Control of a group session is important, as discussion of fire safety issues will trigger not only questions but also comments arising from people's own experience. This is to be encouraged, but if dominant individuals are allowed continuously to develop issues, focus could be lost and the training session may degenerate into open debate. The smaller the group, the easier it is to control such discussions.

Questions and How to Respond

Guidance notes for each of the topics contained in this book are given opposite the relevant page. The content of this should be sufficient to discuss the subject matter and answer general questions. The following gives some recommendations on how to deal with those issues falling outside of the 'straightforward'.

Inevitably, delegates will relate the training to their own workplace and experience, and ask questions of both a general and specific nature. The trainer, if unable to answer authoritatively and immediately, will need to have a response strategy.

For this reason it is recommended that the trainer becomes very familiar with the contents of this book before any sessions are delivered, so that a quick response is given to questions relating to it without the need to 'flick' through the pages. If a question is raised that you cannot answer, note the question down with a promise to reply within a given time frame. This approach should overcome the embarrassing silences or debate that may otherwise occur. Having made the promise to respond, keep it at all costs, as failure will destroy confidence in the system. If possible publish your response widely to the group and beyond.

Frequently Asked Questions (FAQs)

While many of the questions you will be asked – either general or specific – cannot be predicted, experience has shown that certain subjects tend to come up time and time again. To help you deal with these the following FAQs are outlined with suggested standard responses.

Q   If I accept specific duties related to fire safety, what liability do I attract in law?

A   The Fire Authority and, in some cases, the Health and Safety Executive (HSE) have a range of powers under current legislation to prosecute the corporate entity (or where appropriate, an individual) for negligence in respect of fire safety. The prosecution of an individual need not involve a person with specific responsibility for fire safety: it may involve a general member of staff who has acted in a wilfully negligent way, affecting the safety of others. The fact that an employee fails to act in accordance with training given would not normally imply negligence on the part of the trainer.

Q   What is the difference between the requirements of the *Building Regulations* and other fire safety legislation such as the *Fire Precautions Act 1971(FPA 1971)* and the *Fire Precautions (Workplace) Regulations 1997*?

A   The *Building Regulations*, and associated fire safety design guidelines, set the basic fire safety standards for construction or alteration of a building.

Compliance with these standards generally ceases to be monitored once approval has been given and the work carried out – often before the building is occupied. The *FPA 1971* and *Fire Precautions (Workplace) Regulations 1997* seek to ensure on-going fire safety in the occupied building, considering the specific use and risks within the building.

Q   Of all fire safety provisions within buildings, which are the elements that are most often neglected or found to be in need of attention?

A   'Passive' and 'active' fire precautions, as described in **Part 2**, are built-in or fixed items which once installed should need only maintenance and testing. In the case of passive fire compartmentation or separation, elements such as walls and partitions usually require very little maintenance over the lifetime of the building – except perhaps where 'fire-stopping' issues arise due to re-routing of services. From experience of undertaking many fire risk assessments across a wide range of building uses and occupancies, the area of fire precautions most likely to be neglected appears to be fire safety management. This includes many of the topics covered by this manual, such as drawing up effective evacuation procedures and preventing the abuse of self-closing fire doors.

Q   Water sprinklers are an effective response to fire but I am worried about the workplace suffering extensive water damage beyond the range of the fire.

A   Sprinklers are arranged in a grid pattern designed to cover the risk of a developing fire. Only those sprinkler heads affected by a predetermined level of heat exposure will operate. In essence, water will not be released by the system unless a fire is below or adjacent to it. In 95% of fires in buildings fitted with sprinklers, less than three sprinkler heads are activated in controlling a fire.

Q   How do I know if a door is a 'fire door'?

A   Normal fire doors are fitted with a simple self-closing device, and should not be held open (unless they are controlled by an automatic closer linked to the fire alarm system). Any glazing incorporated into a fire door should be fire-resisting (eg Georgian wired glass or Pyran), and modern fire doors should always have smoke seals fitted along their top and side edges.

In most places of work, fire doors will be labelled 'Fire Door – Keep Shut'. However, fire doors leading to cupboards, storerooms or plant rooms are usually kept *locked*, and should be labelled with a sign to this effect. If your building has a fire certificate, this will identify compartment lines on which fire doors should be located.

Q Where is it sensible to create fire evacuation 'Assembly Points'?

A Evacuation from a building should be to one or more Assembly Points clear of the building (ideally at least three times the height of the building away) and clear of any entrance the emergency services will use. Assembly Points should be located so as not to expose people to risk from motor vehicles or any other obvious hazard. If there is no option but to cross roads, a risk assessment must be carried out and careful consideration given to mitigating the risks.

Where possible, Assembly Points should be clearly signposted, and staff made familiar with their location through regular use in fire drills.

Q What are the main conflicts between disabled access law and fire safety evacuation requirements?

A The laws relating to disabled persons' access to buildings and facilities is being strengthened such that it is an offence to discriminate in the provision of employment or services on the basis of disability. This development of the law has led to a rush to provide access facilities for the physically impaired. Sadly, emergency evacuation is not always considered when defining such strategies.

Often, where only normal lifts are installed and most occupants of the building are simply using the stairs, no consideration has been given to how non-ambulant people would evacuate. Disabled people have the same entitlement to means of escape as able-bodied people: for example, where alternative directions of escape exist this level of provision must be available to all users. Arrangements must be made for even-

tual evacuation from the building, possibly involving the identification of 'refuges', obtaining specialist carrying equipment, and training personnel in its use.

Q When should we be worried about arson attacks?

A Your fire risk assessment process should consider if the organisation has any special risks associated with arson. Some of the more obvious considerations would be:

- disgruntled staff (times of staff redundancy or lay-off);

- premises located in volatile neighbourhoods or exposed locations;

- business processes attracting protest group action (animal testing etc);

- high profile premises or institutions that receive close media attention;

- military or establishment premises (armed services, police, customs etc);

- schools; or

- other premises with a history of vandalism or break-ins.

The fire risk assessment should provide for measures to counter the identified risks.

Q Are there any times that my building is at heightened risk from fire?

A Statistics show that one of the largest fire risks to buildings is the introduction of building contractors on site. Contractors by the very nature of the work they undertake often use tools and machinery that generate heat and sparks. The contractors' staff may not be disciplined in your fire precautions, may smoke illicitly and generate unacceptable levels of combustible rubbish in vulnerable areas. Financial pressures on contractors mean that they often 'cut corners' and attempt to save money by sub-contracting services. These tendencies lead to an uncontrolled risk if not managed well.

Q In residential care homes, should any attempt be made to enter a closed room which is clearly well alight, in order to rescue the occupant?

A No. Just opening the door in such circumstances could cause a 'back-draft' situation, producing an explosive fireball that would probably kill both the occupant and their would-be rescuers, and place the entire home in great danger. The best chance of a saving the occupant would be the early attendance of trained and well-equipped firefighters. This emphasises the need to call the fire brigade in response to any actuation of the fire alarm in a care home, except when it is known immediately, and for certain, to be a false alarm.

Record Keeping

Training Attendance

Your organisation must be able to demonstrate that it has planned and executed a fire safety training program. It may be that you will need to produce evidence to this effect, on request, by an inspecting officer of your local fire authority. The only way to ensure this can be done is by keeping clear, detailed training records. A training record sheet should contain some basic information. The fire safety training process may have a hierarchy of courses aimed at persons with differing delegated responsibilities. This structure should be clearly demonstrated in an introduction to the fire safety training record book. **Figure 3.1**, below, demonstrates typical content of a standard training record sheet.

Ideally, the training record will also be kept on the individual's personnel record file – not necessarily in the detail given on the record sheet, but sufficient to indicate that a continuous training program has been undertaken.

The fire safety training record book should be maintained and kept with copies of the Fire Certificate and/or Fire Risk Assessment, fire systems testing and maintenance records, and any other documentation that may be

Name of Course:		Date of Course:	
Name of Trainer:		Time of Course:	
		Duration of Course:	
Delegate Name	Department/ job title	Fire Safety Role	Competency Assessment★
★refer to following section on fire safety training competency assessment			

Figure 3.1: Standard training record sheet

needed for an inspection by the local fire authority. This collection of records usually comprises, or includes, the Fire Log Book.

As a ready record for inspection, fire brigade officers usually prefer hard copy information. However, by negotiation, it may be possible to keep the necessary records on a computer system as long as it is readily accessible and that it can be demonstrated that 'back-up' facilities ensure the integrity of the electronic records.

Training Recognition

To assist in the development of a 'fire safety culture' within an organisation, it has been found by experience that recognition of effort and role goes a long way to gaining participation and involvement. With regard to training, one way of achieving this is to issue training certificates to delegates upon successful completion of a course or series of courses. This is an inexpensive way of recognising what is usually an unrewarded effort.

Employer Organisation and Logo
FIRE SAFETY AWARENESS COURSE
This is to Certify that
Sarah Clark
of
Customer Services
Successfully Completed a Fire Safety Awareness Course on
Date
Mike Williams **Senior Fire Safety Manager**

Figure 3.2: Fire training certificate

The issue of such certificates may be considered as contributing to the organisation's Human Resources management policy, and be reflected in appraisal processes. Whether recognition is formal or informal, it will help to underpin the safety culture in your workplace. An example of typical fire safety training certificates is included in **Figure 3.2.**

Competency Assessment

In **Part 1** of this manual, the advantages and disadvantages of fire safety competency assessment were discussed. It is assumed here that a review of these factors has been undertaken and that the result indicated clear benefits from the use of testing associated with fire safety training. The main advantage of the process is the ability to demonstrate that training has not only been delivered but also understood to a measurable and satisfactory level.

For reasons also discussed in **Part 1**, a full written test is not advocated. Instead, a multiple-choice question and answer sheet is proposed, intended to test understanding not literacy. A sample test sheet follows to indicate possible format and content. This may be set by the trainer as it stands, or adapted to avoid over-familiarity upon repeated use. Further sample questions are then given, including typical questions for specific types of business or workplace.

It is suggested that test papers contain no more than 15 questions. On this basis competency for general staff awareness should be taken as a percentage of correct answers equalling not less than 67% (10 out of 15), though the assessor may wish to give some 'weighting' to certain questions considered more fundamental than others. For those with specific duties, such as Fire Marshals, a higher pass mark should be sought, perhaps 80%, or 12 out of 15, correct. It is suggested that delegates are asked to put aside any course notes or handouts when completing the test. Giving out the correct answers – once all papers have been collected – is a good way of clearing up any misunderstandings before the training session ends and delegates return to work.

If anyone fails to achieve the pass mark required to indicate competency, there are a range of measures that may be considered, from re-taking the course to discussion with the trainer on a one-to-one basis, carefully going through the inappropriately answered questions. What should be avoided at all costs is public discussion of the results attained (other than perhaps some form of commendation for the top achievers). Any comparison that belittles or embarrasses a member of staff in front of colleagues is a recipe for resentment and future non-cooperation.

Sample Fire Safety Awareness Course – Post Course Questionnaire

Course Title here:

Name (Please print clearly):

Organisation/Department: Date:

1   The Fire Precautions (Workplace) Regulations 1997 (as amended) require every employer to:

☐   A. apply for a fire certificate

☐   B. carry out a fire risk assessment

☐   C. install emergency lighting

☐   D. appoint fire wardens

2   By law, the fire risk assessment must be reviewed and if necessary updated:

☐   A. every six months

☐   B. whenever changes occur that may affect the fire risk

☐   C. when instructed by the local Fire Authority

☐   D. annually

3   If you discover a fire at your workplace, your first responsibility is to:

☐   A. fight the fire

☐   B. call the fire brigade

☐   C. raise the alarm

☐   D. report to the assembly area

4   A fire door can usually be recognised by:

☐   A. a sign with a 'running man' symbol

☐   B. a break-glass security device

☐   C. a self-closing device

☐   D. a hook to secure it in the open position

5   A heat detector is more likely to be fitted than a smoke detector in:

☐   A. offices

☐   B. corridors

☐   C. stairways

☐   D. kitchens

6   In fire safety terms the expression 'good housekeeping' refers to:

☐   A. fire drills

☐   B. record keeping

☐   C. improving fire safety through tidiness and keeping the workplace in good order

☐   D. the structural fire safety features of a building

7   The recommended location for the Fire Action Notice is:

☐   A. the staff canteen

☐   B. adjacent to fire alarm points

☐ C. the health and safety manager's office

☐ D. by the fire alarm indicator panel

8   After evacuating the premises staff should:

☐ A. go home and await further instructions

☐ B. return to work as soon as they think it is safe to do so

☐ C. make phone calls to arrange transport

☐ D. report to the correct assembly area

9   The correct colour coding for a dry powder extinguisher is:

☐ A. cream

☐ B. green

☐ C. blue

☐ D. red

10   An extinguisher bearing a black label should contain:

☐ A. foam

☐ B. water

☐ C. wet chemical

☐ D. carbon dioxide

11   A water extinguisher is suitable for fires involving:

☐ A. a spillage of burning fat

☐ B. live electrical equipment

☐ C. escaping gas

☐ D. wood, cardboard and textiles

12   Before attacking a fire with an extinguisher you should:

☐ A. test the extinguisher by discharging it momentarily

☐ B. put on protective goggles and gloves

☐ C. open some windows to let the smoke out

☐ D. close the door behind you

13   A potential arsonist is likely to be deterred by:

☐ A. a secluded yard hidden by a tall hedge

☐ B. skips full of waste paper

☐ C. closed circuit television

☐ D. an unmanned security post

14   It is acceptable for a fire exit to open from within by means of:

☐ A. a panic bolt

☐ B. a removable key

☐ C. a security code

☐ D. a swipe card

15   If your fire alarm system calls out to the fire brigade automatically, you should always:

☐ A. search the building for the cause of the alarm

☐ B. make a follow-up call to the fire brigade

☐ C. do nothing more

☐ D. wait for the fire brigade to arrive before evacuating the building

Additional Questions

These questions may be substituted into the sample questionnaire given above in order to vary the presentation and to build a wider sphere of understanding. As more questions are required it is suggested that the fire safety trainer develops their own 'bank' of questions, varying the level of difficulty, by using the text of **Part 2** and experience of their own workplace: its layout, facilities, workforce and problems. If this advice is followed, both the trainer and the delegates will develop a wider understanding of the risks in their premises.

1   An extinguisher bearing a yellow label should contain:

☐   A. foam

☐   B. water

☐   C. wet chemical

☐   D. carbon dioxide

2   When fighting a fire with a hosereel you should continue:

☐   A. for twenty minutes

☐   B. for not more than two or three minutes

☐   C. until the fire is out

☐   D. until the fire brigade arrives

3   The only acceptable method of holding open a fire door is:

☐   A. an automatic closing device linked to the fire alarm system

☐   B. a wooden wedge

☐   C. a retaining hook fixed to the wall

☐   D. a fire extinguisher

4   When laundry has finished drying in a tumble drier, it should be:

☐   A. left in the drier and allowed to cool down

☐   B. unloaded immediately, shaken out and folded

☐   C. given a further ten minutes drying time

☐   D. unloaded but not separated until convenient to do so

5   To prevent fire spreading to the building, rubbish bins should be kept:

☐   A. securely locked at all times

☐   B. away from overhanging branches

☐   C. close to the back door

☐   D. at least ten metres from any external wall

6   In a care home for frail or elderly people, when the fire alarm sounds residents should be:

☐   A. evacuated to the car park

☐   B. encouraged to help each other down the stairs

☐   C. allowed to stay in their rooms, if they are in no danger

☐   D. told it doesn't concern them

7   'Horizontal evacuation' in care home or healthcare premises will always involve:

☐   A. crawling through smoke-filled passageways

☐   B. passing through at least one set of fire doors and allowing them to close behind you

☐   C. walking down one flight of stairs

☐   D. helping all residents out of the building

8   If care home residents are thought to be in danger from a fire, they should if possible be:

☐ A. told to stay in their rooms

☐ B. instructed to leave the building immediately

☐ C. left for the fire brigade to deal with

☐ D. helped to move into a protected area on the same floor

9 'Maintained' emergency lighting is lighting which:

☐ A. is regularly serviced

☐ B. only comes on when the normal lighting fails

☐ C. is used normally and stays on when the normal lighting fails

☐ D. is left on 24 hours a day

10 In the event of the fire alarm sounding, lifts can be used:

☐ A. if they are disabled evacuation lifts controlled by staff

☐ B. by members of staff only

☐ C. in areas not apparently affected by the fire

☐ D. under no circumstances whatsoever

11 In large buildings with a phased evacuation procedure, people who wish to evacuate on hearing the 'alert' signal:

☐ A. should be discouraged from evacuating but not restricted

☐ B. should be prevented at all costs

☐ C. should be reported to security

☐ D. should be accompanied to the Assembly Point

12 Fire drills should be carried out:

☐ A. weekly when testing the fire alarm

☐ B. twice yearly, or more often if required by the Fire Authority or the standards laid down for your industry

☐ C. outside of normal working hours

☐ D. monthly

13 'Risk' is quantified by multiplying together **(tick <u>two</u> boxes)**

☐ A. the likelihood of a hazard causing harm

☐ B. the number of people working in the vicinity

☐ C. the potential severity of the consequences

☐ D. the control measures in place

14 The control measures to be considered when assessing a fire risk include:

☐ A. applicable fire safety legislation

☐ B. adequate insurance cover

☐ C. presence of staff in the area of the risk

☐ D. a current fire certificate

15 In an evacuation, the Assembly Point Officer or Marshal is responsible for:

☐ A. calling the Fire Brigade

☐ B. organising the Roll Call and reporting on evacuation status

☐ C. directing fire-fighting activities

☐ D. searching the building for missing people

Sample Fire Safety Awareness Course – Answers to Post Course Questionnaire

1 The Fire Precautions (Workplace) Regulations 1997 (as amended) require every employer to:

☐   B. carry out a fire risk assessment

2   By law, the fire risk assessment must be reviewed and if necessary updated:

☐   B. whenever changes occur that may affect the fire risk

3   If you discover a fire at your workplace, your first responsibility is to:

☐   C. raise the alarm

4   A fire door can usually be recognised by:

☐   C. a self-closing device

5   A heat detector is more likely to be fitted than a smoke detector in:

☐   D. kitchens

6   In fire safety terms the expression 'good housekeeping' refers to:

☐   C. improving fire safety through tidiness and keeping the workplace in good order

7   The recommended location for the Fire Action Notice is:

☐   B. adjacent to fire alarm points

8   After evacuating the premises staff should:

☐   D. report to the correct assembly area

9   The correct colour coding for a dry powder extinguisher is:

☐   C. blue

10   An extinguisher bearing a black label should contain:

☐   D. carbon dioxide

11   A water extinguisher is suitable for fires involving:

☐   D. wood, cardboard and textiles

12   Before attacking a fire with an extinguisher you should:

☐   A. test the extinguisher by discharging it momentarily

13   A potential arsonist is likely to be deterred by:

☐   C. closed circuit television

14   It is acceptable for a fire exit to open from within by means of:

☐   A. a panic bolt

15   If your fire alarm system calls out to the fire brigade automatically, you should always:

☐   B. make a follow-up call to the fire brigade

Answers to Additional Questions

1   An extinguisher bearing a yellow label should contain:

☐   C. wet chemical

2   When fighting a fire with a hosereel you should continue:

☐   B. for not more than two or three minutes

3   The only acceptable method of holding open a fire door is:

☐   A. an automatic closing device linked to the fire alarm system

4   When laundry has finished drying in a tumble drier, it should be:

☐   B. unloaded immediately, shaken out and folded

5   To prevent fire spreading to the building, rubbish bins should be kept:

☐   D. at least ten metres from any external wall

6   In a care home for frail or elderly people, when the fire alarm sounds residents should be:

☐   C. allowed to stay in their rooms, if they are in no danger

7   'Horizontal evacuation' in care home or healthcare premises will always involve:

☐ B. passing through at least one set of fire doors and allowing them to close behind you

8 If care home residents are thought to be in danger from a fire, they should if possible be:

☐ D. helped to move into a protected area on the same floor

9 'Maintained' emergency lighting is lighting which:

☐ C. is used normally and stays on when the normal lighting fails

10 In the event of the fire alarm sounding, lifts can be used:

☐ A. if they are disabled evacuation lifts controlled by staff

11 In large buildings with a phased evacuation procedure, people who wish to evacuate on hearing the 'alert' signal:

☐ A. should be discouraged from evacuating but not restricted

12 Fire drills should be carried out:

☐ B. twice yearly, or more often if required by the Fire Authority or the standards laid down for your industry

13 'Risk' is quantified by multiplying together **(tick <u>two</u> boxes)**

☐ A. the likelihood of a hazard causing harm

☐ C. the potential severity of the consequences

14 The control measures to be considered when assessing a fire risk include:

☐ C. presence of staff in the area of the risk

15 In an evacuation, the Assembly Point Officer or Marshal is responsible for:

☐ B. organising the Roll Call and reporting on evacuation status

Notes

Notes

Notes

Notes

Notes

Notes